ברכי נפשי

THE WONDERS OF CREATION

MARVELS OF OUR WORLD

MOSAICA PRESS

Mosaica Press, Inc.

Designed by Rayzel Broyde
Typeset by Daniella Kirsch

ISBN-10: 1-946351-49-0
ISBN-13: 978-1-946351-49-4

Published by:
Mosaica Press, Inc.
www.mosaicapress.com
info@mosaicapress.com

THE WONDERS OF CREATION

MARVELS OF OUR WORLD

RABBI YAAKOV LUBIN

THIS BOOK IS DEDICATED TO

Andrew and Samantha Sossin

Their friendship and support has enabled this book to be written and published I would not be where I am today, publishing this book and running The Jewish Center for Science, without Andrew's constant encouragement and support. Andrew has always supported me in my endeavors, *l'shem Shamayim*, and I cannot begin to express my gratitude for how he has helped enable me to make sanctifying Hashem's name my pursuit in life. May Andrew, Samantha, and their children be blessed with health, happiness, and prosperity.

A SPECIAL THANKS GOES TO

Grant and Sandra Einhorn

for their support of and dedication to The Jewish Center for Science. Thank you for being great friends and all of the work you have put in to seeing this project succeed.

הרב שמעון גלאי

רב ומו"צ

רח' אהרונוביץ 51 בני ברק

בס"ד, תאריך: ז' אדר ע"ז

מכתב ברכה

הריני לחזק בזה את ידי הרב יעקב לוין שליט"א שלקח
על עצמו להדפיס את המובא בחובת הלבבות
להתבונן בחסדי השי"ת שרואים מתוך הבריאה
ואחד מהדברים שצריכים חיזוק בכלל הוא
דבר זה, להתבונן בחסדיו בכל עת ובכל שעה
כאמור "ואני ברב חסדך אבוא ביתך"
ומצוה גדולה לחזק אותם בכל מה שאפשר
שיצליחו להחדיר ענין זה בכל תפוצות ישראל
וכן ראיתי לחזק ידי הרב משה הירש שליט"א שמסייע
מאוד בענין זה

החותם בברכה
שמעון גלאי

בס"ד

מכתב המלצה

הנה מודעת זאת מפי ספרים ומפי סופרים ורבותינו מצוקי ארץ הגידו לנו מראש את אשר יקרה באחרית הימים. כאשר החושך יכסה ארץ, ושלהבת אור האמונה המושרשת בלב כל זרע אברהם אשר בשם ישראל יכונה- שבשנים כתיקונם היתה עולה מאליה ועוברת בירושה מדור לדור בכל תוקף זיוה והדרה- בעקבתא דמשיחא תהא נעדרת וטמונה בין הרי החושך, ורק ע"י יגיעה ועבודה מרובה ניתן לדלותה ולגלותה ממסתריה ממעמקי הלב אשר שם יתלחשון גחליה.

ועל כן בואו ונחזיק טיבותא לידידי וו"ח יראת ה' היא אוצרו הרה"ח ר' יעקב לובין שליט"א אשר רחש לבו הטהור דבר טוב לחזק ברכים כושלות, ולמטרה זו יסד מכון "ברכי נפשי" למען הקנות לצעירי הצאן את ערכי אמונה הטהורה ע"י הרחבת והמחשת בביאור מעשי ה' ונפלאותיו הסובבים אותנו תמיד, ובחכמת ה' וחסדיו המרובים אשר משפיע לבריאותיו בכל עת. וכידוע מרבותינו הרמב"ם ובעל חובת הלבבות ז"ל ועוד, כמה כוחו רב לגלות ולהבעיר את האמונה הטמונה ולהביא לידי אהבת ה' ויראתו.

והנני מברכו שחפץ ה' בידו יצליח לגדל עדרים עדרים של צאן קדשים הרועים אמונה ומתעדנים מנאות דשאיה ומתיקות פירותיה, ובזכות האמונה נזכה במהרה לאור חדש על ציון ואז נשיר שיר חדש מראש אמנה על גאולתינו ועל פדות נפשינו

משה וואלפסאן

TEHILLIM 104: BARCHI NAFSHI

1 My soul, bless the Lord. My God, You are very great, You are attired with majesty and beauty.

א בָּרְכִי נַפְשִׁי אֶת־יְהֹוָה יְהֹוָה
אֱלֹהַי גָּדַלְתָּ מְּאֹד הוֹד וְהָדָר לָבָשְׁתָּ:

2 [You] enwrap Yourself with light like a garment; [You] extend the heavens like a curtain.

ב עֹטֶה אוֹר כַּשַּׂלְמָה נוֹטֶה שָׁמַיִם
כַּיְרִיעָה:

3 Who roofs His upper chambers with water; Who makes clouds His chariot, which goes on the wings of the wind.

ג הַמְקָרֶה בַמַּיִם עֲלִיּוֹתָיו הַשָּׂם־
עָבִים רְכוּבוֹ הַמְהַלֵּךְ עַל־כַּנְפֵי־
רוּחַ:

4 He makes winds His messengers, burning fire His ministers.

ד עֹשֶׂה מַלְאָכָיו רוּחוֹת מְשָׁרְתָיו
אֵשׁ לֹהֵט:

5 He founded the earth on its foundations that it not falter to eternity.

ה יָסַד אֶרֶץ עַל־מְכוֹנֶיהָ בַּל־תִּמּוֹט
עוֹלָם וָעֶד:

6 You covered the deep as [with] a garment; the waters stand on the mountains.

ו תְּהוֹם כַּלְּבוּשׁ כִּסִּיתוֹ עַל־הָרִים
יַעַמְדוּ מָיִם:

7 From Your rebuke they fled; from the sound of Your thunder they hastened away.

ז מִן־גַּעֲרָתְךָ יְנוּסוּן מִן־קוֹל רַעַמְךָ
יֵחָפֵזוּן:

8 They ascended mountains, they descended into valleys to this place, which You had founded for them.	**ח** יַעֲלוּ הָרִים יֵרְדוּ בְקָעוֹת אֶל-מְקוֹם זֶה | יָסַדְתָּ לָהֶם:
9 You set a boundary that they should not cross, that they should not return to cover the earth.	**ט** גְּבוּל-שַׂמְתָּ בַּל-יַעֲבֹרוּן בַּל-יְשׁוּבוּן לְכַסּוֹת הָאָרֶץ:
10 He sends the springs into the streams; they go between the mountains.	**י** הַמְשַׁלֵּחַ מַעְיָנִים בַּנְּחָלִים בֵּין הָרִים יְהַלֵּכוּן:
11 They water every beast of the field; the wild donkeys quench their thirst.	**יא** יַשְׁקוּ כָּל-חַיְתוֹ שָׂדָי יִשְׁבְּרוּ פְרָאִים צְמָאָם:
12 Beside them the fowl of the heavens dwell; from between the branches they let out their voices.	**יב** עֲלֵיהֶם עוֹף-הַשָּׁמַיִם יִשְׁכּוֹן מִבֵּין עֳפָאיִם יִתְּנוּ-קוֹל:
13 He waters the mountains from His upper chambers; from the fruit of Your works the earth is sated.	**יג** מַשְׁקֶה הָרִים מֵעֲלִיּוֹתָיו מִפְּרִי מַעֲשֶׂיךָ תִּשְׂבַּע הָאָרֶץ:
14 He causes grass to sprout for the animals and vegetation for the work of man, to bring forth bread from the earth.	**יד** מַצְמִיחַ חָצִיר | לַבְּהֵמָה וְעֵשֶׂב לַעֲבֹדַת הָאָדָם לְהוֹצִיא לֶחֶם מִן-הָאָרֶץ:
15 And wine, which cheers man's heart, to make the face shine from oil, and bread, which sustains man's heart.	**טו** וְיַיִן | יְשַׂמַּח לְבַב-אֱנוֹשׁ לְהַצְהִיל פָּנִים מִשָּׁמֶן וְלֶחֶם לְבַב-אֱנוֹשׁ יִסְעָד:

16 The Lord's trees are sated, the cedars of Lebanon, which He planted.

טז יִשְׂבְּעוּ עֲצֵי יְהֹוָה אַרְזֵי לְבָנוֹן אֲשֶׁר נָטָע׃

17 Where birds nest; as for the stork, the high junipers are its home.

יז אֲשֶׁר־שָׁם צִפֳּרִים יְקַנֵּנוּ חֲסִידָה בְּרוֹשִׁים בֵּיתָהּ׃

18 The high mountains for the wild goats and the rocks as refuge for the gofers.

יח הָרִים הַגְּבֹהִים לַיְּעֵלִים סְלָעִים מַחְסֶה לַשְׁפַנִּים׃

19 He made the moon for the appointed seasons; the sun knows its setting.

יט עָשָׂה יָרֵחַ לְמוֹעֲדִים שֶׁמֶשׁ יָדַע מְבוֹאוֹ׃

20 You make darkness and it is night, in which every beast of the forest moves about.

כ תָּשֶׁת חֹשֶׁךְ וִיהִי לָיְלָה בּוֹ־תִרְמֹשׂ כָּל־חַיְתוֹ־יָעַר׃

21 The young lions roar for prey and to beg their food from God.

כא הַכְּפִירִים שֹׁאֲגִים לַטָּרֶף וּלְבַקֵּשׁ מֵאֵל אָכְלָם׃

22 When the sun rises they gather in and crouch in their dens.

כב תִּזְרַח הַשֶּׁמֶשׁ יֵאָסֵפוּן וְאֶל־מְעוֹנֹתָם יִרְבָּצוּן׃

23 Man goes out to his work, to his labor until evening.

כג יֵצֵא אָדָם לְפָעֳלוֹ וְלַעֲבֹדָתוֹ עֲדֵי־עָרֶב׃

24 How great are Your works, O Lord! You have made them all with wisdom; the earth is full of Your possessions!

כד מָה־רַבּוּ מַעֲשֶׂיךָ | יְהֹוָה כֻּלָּם בְּחָכְמָה עָשִׂיתָ מָלְאָה הָאָרֶץ קִנְיָנֶךָ׃

25 This sea, great and wide; there are creeping things and innumerable beasts, both small and large.

כה זֶה | הַיָּם גָּדוֹל וּרְחַב יָדָיִם
שָׁם רֶמֶשׂ וְאֵין מִסְפָּר חַיּוֹת קְטַנּוֹת
עִם־גְּדֹלוֹת:

26 There the ships go; You formed this leviathan with which to sport.

כו שָׁם אֳנִיּוֹת יְהַלֵּכוּן לִוְיָתָן זֶה |
יָצַרְתָּ לְשַׂחֶק־בּוֹ:

27 They all look to You with hope, to give their food in its time.

כז כֻּלָּם אֵלֶיךָ יְשַׂבֵּרוּן לָתֵת
אָכְלָם בְּעִתּוֹ:

28 You give them that they may gather; You open Your hand that they may be sated with goodness.

כח תִּתֵּן לָהֶם יִלְקֹטוּן תִּפְתַּח יָדְךָ
יִשְׂבְּעוּן טוֹב:

29 You hide Your countenance and they are frightened; You gather in their spirit and they perish and return to their dust.

כט תַּסְתִּיר פָּנֶיךָ יִבָּהֵלוּן תֹּסֵף
רוּחָם יִגְוָעוּן וְאֶל־עֲפָרָם יְשׁוּבוּן:

30 You will send forth Your spirit and they will be created, and You will renew the surface of the ground.

ל תְּשַׁלַּח רוּחֲךָ יִבָּרֵאוּן וּתְחַדֵּשׁ
פְּנֵי אֲדָמָה:

31 The glory of the Lord will be forever; the Lord will rejoice with His works.

לא יְהִי כְבוֹד יְהוָה לְעוֹלָם יִשְׂמַח
יְהוָה בְּמַעֲשָׂיו:

32 He Who looks at the earth and it quakes; He touches the mountains and they emit smoke.

לב הַמַּבִּיט לָאָרֶץ וַתִּרְעָד יִגַּע
בֶּהָרִים וְיֶעֱשָׁנוּ:

33 I shall sing to the Lord while I am alive; I shall sing praises to my God as long as I exist.

34 May my speech be pleasing to Him; I shall rejoice with the Lord.

35 Sinners will be destroyed from the earth and the wicked will be no more; my soul, bless the Lord. Hallelujah.

לג אָשִׁירָה לַיהֹוָה בְּחַיָּי אֲזַמְּרָה לֵאלֹהַי בְּעוֹדִי:

לד יֶעֱרַב עָלָיו שִׂיחִי אָנֹכִי אֶשְׂמַח בַּיהֹוָה:

לה יִתַּמּוּ חַטָּאִים | מִן־הָאָרֶץ וּרְשָׁעִים | עוֹד אֵינָם בָּרְכִי נַפְשִׁי אֶת־יְהֹוָה הַלְלוּיָהּ:

TABLE OF CONTENTS

ACKNOWLEDGMENTS

I want to thank Hashem for giving me the *zechus* to both learn and teach about the amazing world that He made.

There have been many people who helped make this book possible and I would like to give recognition to some of them. May all those who contributed to the completion of this book and creation of The Jewish Center for Science be blessed with much success for their help.

The first person I want to thank is my partner and co-founder of The Jewish Center for Science, Andrew Sossin, to whom this book is dedicated and whose friendship and dedication have been truly incredible. I would not be who I am or where I am today if not for him. He has been an inspiration to me and many others. May Hashem give him the success and strength to continue supporting my projects and the others he is involved with.

Grant Einhorn has been instrumental in starting The Jewish Center for Science, and all his help is greatly appreciated and worthy of praise.

My rabbi, Rabbi Shmuel Rosenbaum, *shlita*, has given me great support in pursuing the study and teaching of *nifla'os haBorei*. His encouragement and guidance have helped me for many years and have played a fundamental role in my learning Torah, writing this book, and in creating The Jewish Center for Science.

My thanks go to *Mishpacha* magazine, which allowed us to use the contents of the series I wrote, "Marvel of the Month," in this book.

It is a special privilege to thank my parents and family for all they have done and continue to do for me. May they get much *nachas* from this book and the other projects I am involved with.

Finally, I want to thank my *ezer k'negdo*, Miri, whom I have been fortunate to share my life with and who has been instrumental in the making of this book and The

Jewish Center for Science. She is the "behind the scenes" in everything I do. Her help with editing, encouragement, and support in every way has been essential for this book being completed and everything else I have and will in the future accomplish.

INTRODUCTION

Why did Hashem make such a beautiful world? The Creator in His infinite wisdom could have made a much simpler world without sunsets, mountains, flowers, and trees.

One of the reasons why Hashem made a world with the multitudes of wonders we have the privilege to see and enjoy is so that we could recognize His greatness and glory; to thank and praise Him for His infinite wisdom and kindness. As we reflect upon the world around us, it becomes a powerful means to connect us to our Creator in a deeper and more meaningful way.

The great Torah sage Rabbi Moshe ben Maimon, known as the Rambam, tells us in his code of Jewish law how we are to properly relate to the wonders that we see around us. One of the mitzvos in the Torah is to love G-d. You may ask: How can one be obligated to feel something? What place does something like love have in a book of laws?

The Rambam instructs us on how to fulfill the mitzvah of loving G-d in the section of his *Mishneh Torah* that contains the laws that are the foundations of the Torah (chapter 2, law 2), which states:

> *What is the path to come to a love and fear of G-d? When a person contemplates His great and wondrous creations and deeds and sees from them His wisdom which has no comparison or end, he will immediately love, praise, and glorify [Him], yearning with tremendous desire to know [G-d's] great name, as King David stated: "My soul thirsts for the Lord, for the living G-d" (Tehillim 42:3).*

The Rambam is telling us that the more we recognize of G-d and His deeds, the greater our love and awe of Him will be. While we cannot be forced to feel love, we can be instructed to perform actions that will generate such feelings.

Despite the fact that reflecting on the wonders of the Creator is so important to our relationship with Him, many of us fail to do it very often. In the classic work *Duties of the Heart*, Rabbeinu Bachya dedicates one of the ten gates toward teaching how to properly reflect upon the wonders of Hashem. There he gives several reasons why we tend not to notice the awesome and amazing creations all around us:

- We are too busy chasing after things like money and success.
- We get used to seeing things when we are young and have undeveloped minds. When we grow up and can really start to appreciate what amazing gifts Hashem gave us, we don't notice them because we are so familiar with them.
- The last reason he gives is because of the suffering and pain we go through in life that can distract us from seeing what is good and beautiful in the world.

In order to combat these pitfalls that prevent us from seeing the abundant wisdom and kindness of Hashem, we need to designate time to reflect. A small amount of effort in this area can bring tremendous rewards as there are endless amounts of gifts from Hashem we have that we don't even know about because we never stopped to notice them.

As we explore the *nifla'os Hashem* we will begin to understand a little better what David HaMelech meant in *Tehillim* 104 when he said, "*Mah rabu maasecha Hashem kulam b'chochmah asisa*." Remember, however, that this book is not meant to be the final destination of exploring the wonders of the Creator but the beginning of a life of heightened awareness of the favors that Hashem does for us and the unfathomable wisdom and kindness that exist in everything He made.

CHAPTER 1

WONDROUS WATER

> *The purpose of all these wonders is to make us aware that Hashem is there and that He is the One who brought these wonders about. That is the only reason for the brilliant design, engineering, and logistics that we see in everything.*
>
> *Rabbi Avigdor Miller (Ohr Avigdor, introduction to Chovos HaLevavos)*

Did you drink a glass of water today, wash your hands, or go swimming in a pool? You may not have realized it, but you were coming in direct contact with one of the most marvelous molecules in the universe! In fact, it's the unique properties of water that enable all life on our planet to exist.

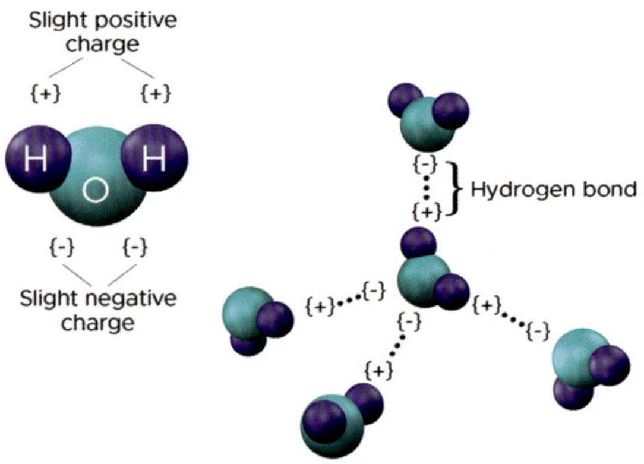

You can get an idea of how water molecules "stick" to each other like mini-magnets in this diagram, where you see the positively charged hydrogen atoms forming short-lived bonds, called hydrogen bonds, with the negatively charged oxygen atoms of their neighbors.

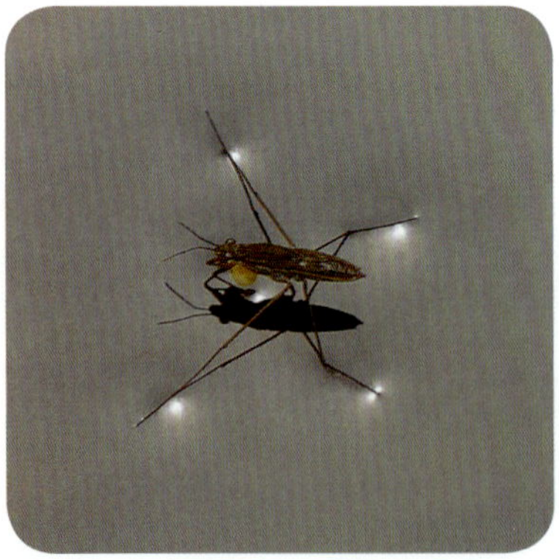

It is the binding of water molecules that gives these water striders a "solid" surface to walk on.

The secret to water's success lies in its exquisite design. Water molecules are made in such a way that they stick to one another. If you don't believe me, try filling a cup of water a little bit above the rim. The reason it doesn't spill over is because the individual water molecules pull on each other in very weak and short-lived bonds that scientists call **HYDROGEN BONDS**. Although each bond is frail and doesn't last long, at any given time most of the water molecules are holding onto each other, preventing the water from spilling over.

Water molecules don't just form bonds with each other, they actually bind to just about everything they come into contact with. When something like salt falls into water, the water molecules grab onto and pull apart the individual atoms the salt is made up of. In a short amount of time, water can dissolve many of the substances needed for life. If not for water, the nutrients you need to stay healthy would be too big for the microscopic cells of your body to use. Water helps break them down, making them useful and easily transportable to all parts of your body. In fact, water's ability to break other substances down is so remarkable that scientists refer to it as the universal solvent. A solvent is a substance that can break down other things and water is better at doing this than anything else on earth.

Flowing rivers break down rocks and soil, providing necessary minerals for plants and animals to drink.

Another unique property of water is that it floats when it freezes. Every liquid other than water gets denser and heavier upon freezing, which makes them sink. If water were to

In addition to keeping the fish warm, frozen lakes can be lots of fun!

"follow the crowd" and sink when it froze, most oceans and lakes would freeze solid in the winter and the countless creatures that live in them would die from the cold. Hashem didn't want that to happen so He made it that when water freezes, the bonds between water molecules lock into place and the molecules are kept farther away from each other than when in liquid form. This results in ice having more empty space between molecules than water, making it lighter and able to float. The resulting layer of ice on top of oceans and lakes also helps keep the fish warm by forming a sort of insulating blanket that keeps out the bitter cold of the winter. In other words, the ice on top keeps the bottom cozy!

If not for the "sticky" nature of water, these trees wouldn't make it more than a few inches off the ground.

Enabling trees to grow tall is surely another wonder of water. Trees need a way to transport the water from underneath the ground to the branches and leaves at their heights. This is no simple task because the force of gravity is always pulling the water down toward the ground. The major force in water transport within trees is provided by the evaporation of water from leaves. Since water molecules are stuck to each other, when water moves out of a leaf, it pulls the rest of the water up along with it. This process is known as **TRANSPIRATION PULL** and can get water up to the top of even the tallest trees.

Isn't it amazing how Hashem uses the seemingly simple water molecule to do so many things for the plants, animals, and people of His world! If you want to see more of the kindnesses of your Creator, try thinking of all the ways you personally benefit from water. You may be surprised to find just how useful water is to you.

CHOCHMAH V'CHESSED

Many people refer to earth as the "blue planet," and this is due to the abundance of water that enables life on our planet to exist.

If we take some time to reflect upon it, the wisdom we find in Hashem's design of water is almost unbelievable. From a simple water molecule, amazing properties arise. It can break down into microscopic pieces and transport all of the salts, minerals, and nutrients needed for us to survive. Water gets lighter upon freezing — an ability that is not found with any other substance — which prevents the oceans from freezing solid. Within trees, water molecules stick together with enough strength to form water ropes that can lift it to the tops of the tallest trees. And as we will see in the upcoming chapters, it is extremely easy to transport. With a little bit of sunshine, it evaporates into the air, making it easy to send around the world where it helps keep us cool and quenches the thirst of creatures around the globe.

While the Divine wisdom found in water is truly great, the kindness of water is no less impressive. Whether it's for swimming pools, washing the floor, keeping us clean, quenching our thirst, or cooking dinner, water plays a central role in making our lives easier and more enjoyable. Just think, each and every time we come in contact with water, we are encountering a spectacular example of Hashem's Divine design.

CHAPTER
2

MASHIV HARUACH

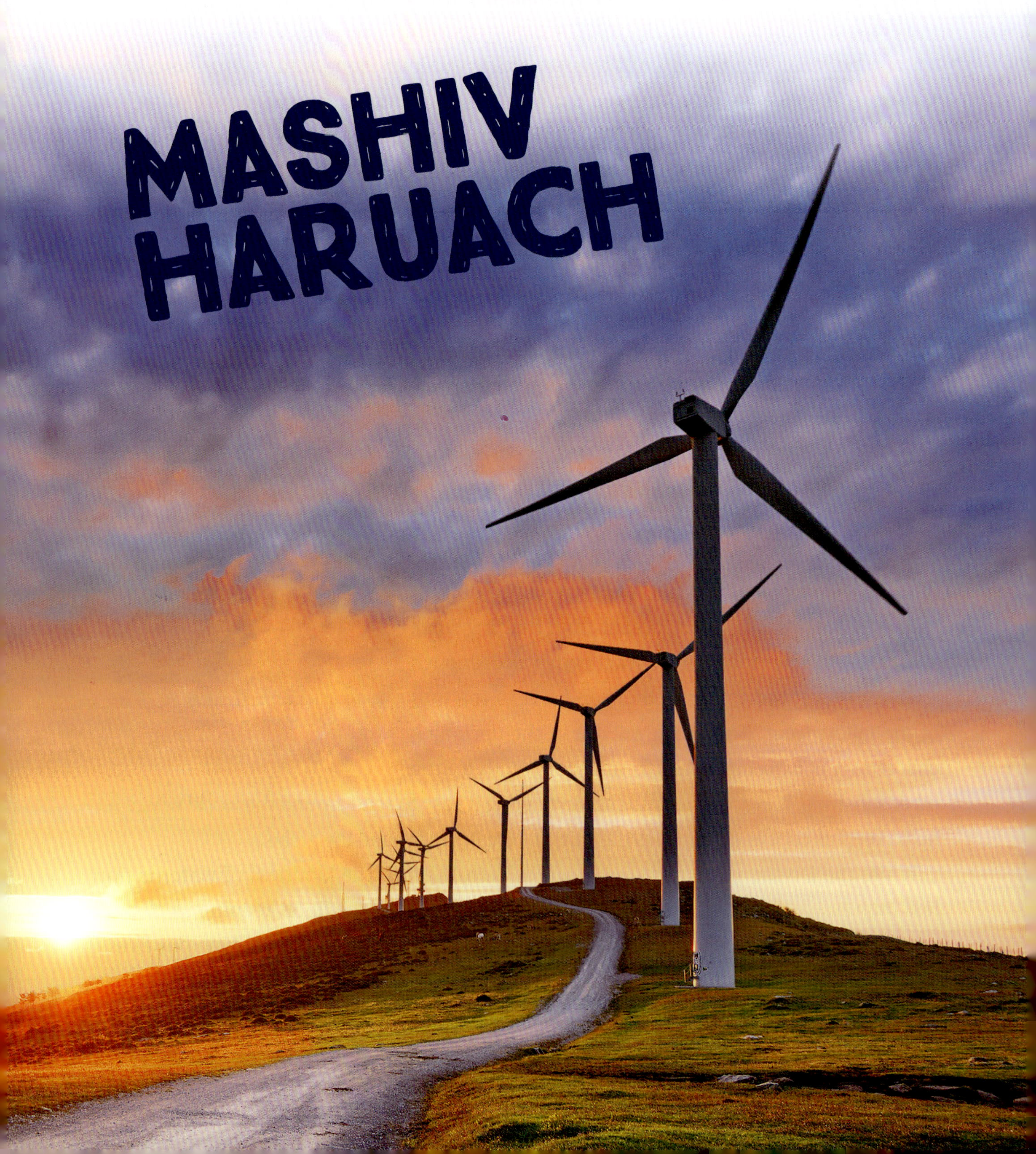

Raise up your eyes and see who created all these.
(Yeshayah 40:26)

Imagine there was a worldwide effort to design a system to distribute billions of tons of fresh water around the world. It would run continuously, have no electric bills, never break down or need repairs...If you think it sounds impossible, you're right! Even the smartest of people could never accomplish such a feat.

Hashem, on the other hand, can do anything! He created an ingenious water-distribution system for our world, using wind to transfer water from the oceans where it is abundant to the land where it is not.

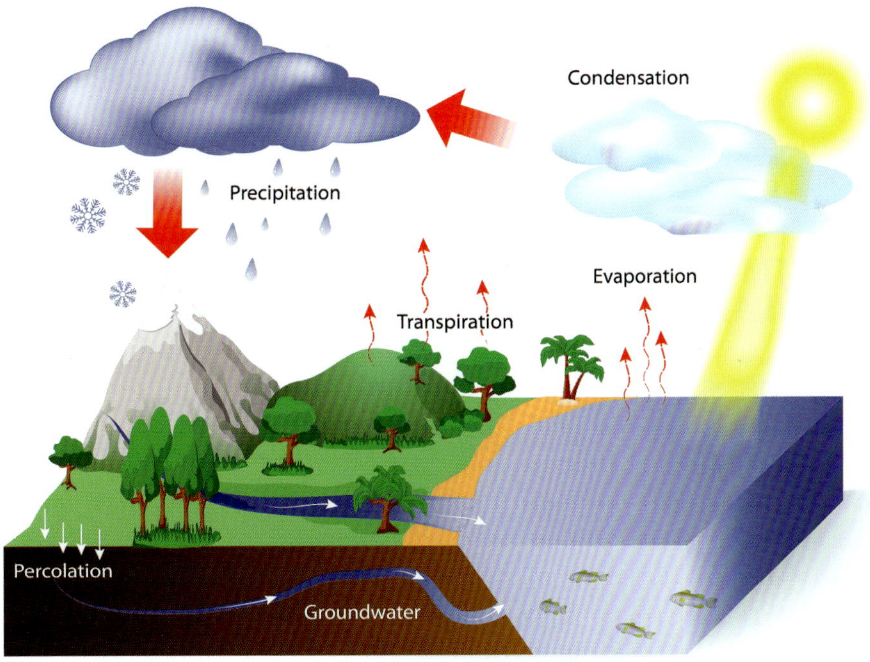

Air has an amazing ability to carry water and help it make its move from the oceans to us by way of the wind.

Water's journey from the sea begins near the equator, where an abundance of sunlight causes massive amounts of it to evaporate into the air. Just like steam rising from a kettle,

It is the rising hot air in hot-air balloons that takes people for rides in the sky!

water rises from the surface of the water that was heated by the sun. It isn't just water that's on the rise; all the air near the earth's surface is heated and ascends into the sky.

Once the air rises high into the sky, it picks up speed, forming the **JET STREAMS**, which are rapidly flowing currents of air that pilots use to make their planes go faster and your flight shorter. The air then travels about 2,100 miles north and south of the equator, until it cools down and begins its descent. When the cold air comes crashing down, it has nowhere else to go so it is forced both toward and away from the equator, forming steady flowing winds near the earth's

GLOBAL ATMOSPHERIC CIRCULATION

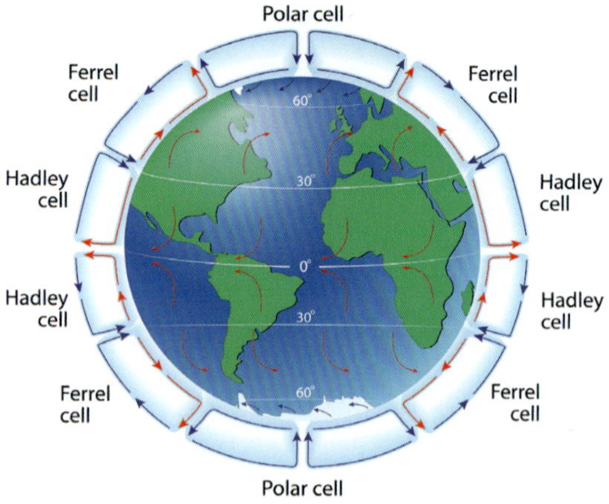

Here is a diagram of how Hashem moves air around our world. The red arrows represent the warm air that rises high into the sky, gets carried by the jet streams, and then falls back to the surface after it's cooled down.

Sailing would be rather boring without the wind.

surface. In all, the air that rose near the equator goes in a big loop, taking it a third of the way around the world and then back again—all in about nine days.

There are actually three large loops of circulating air that combine to carry water around the world and produce the steady, predictable winds of our planet. Stable wind patterns are truly a gift from Hashem. They have benefited mankind for thousands of years by enabling ships to sail around the world. With reliable winds, sailors can catch a ride with a wind that will take them to another country and then ride a different one back when they are ready to go home.

With our newfound knowledge, we can better understand why Hashem called the sky "*shamayim*." Rashi teaches us that the name *shamayim* can be read *aish mayim*, *sa mayim*, or *sham mayim* (*Bereishis* 1:8). This is because the heat from the sun (*aish*) causes the water from the oceans to evaporate into the air. The water is then lifted up (*sa*) until it cools and turns into clouds that are *sham* (up there), laden with *mayim*.

CHOCHMAH V'CHESSED

Here we discover the Divine wisdom found in wind. Wind is an incredibly efficient worldwide water transportation system that delivers twenty trillion gallons of fresh water around the world every year. Driven by the heat of the sun, predictable wind currents also enabled world travel and trade long before airplanes. When we think about all the additional benefits we get from wind (such as keeping us cool, generating electricity, and flying kites, to name just a few), the statement "*Milo kol ha'aretz kevodo*" takes on new meaning as the glory of Hashem we find in the wind is literally found everywhere on earth.

Can you believe wind is strong enough to provide electricity for thousands of homes, through wind turbines like these?

Wind can even be lots of fun to play with!

CHAPTER
3

U'MORID HAGESHEM

The rabbis taught: On the occasion of every distress that befalls the public, we sound the alarm [i.e., we call for prayer and fasting]—with the exception of excessive rain. Why? Rabbi Yochanan says, "Because we do not offer supplication over an abundance of good."

(Taanis 22b)

Have you ever seen a freight train flying in the sky? I like to think of the clouds passing over my head as freight trains carrying their precious cargo of water to thirsty creatures around the world. While you've probably spent some time marveling at the impressive sights above your head, have you ever wondered how Hashem makes clouds and rain?

It all starts from the "seeds" that clouds grow from, which are teeny-tiny pieces of "stuff" high up in the sky. They are referred to as **CLOUD SEEDS** and come from many places including giant sandstorms, volcanic eruptions, and car exhaust that makes its way up to where clouds are formed. Just like any seed, to grow into clouds these need water.

With the booming blast of a volcano, billions of cloud seeds get sent around the world.

There is usually enough water in the air to make clouds because water evaporates from oceans, trees, and other bodies of water and is distributed around the world by way of the wind. The reason why clouds don't always form is because the air is often too hot: With hot air, the water molecules are moving around too fast to hit the cloud seeds. When the air cools down, the water molecules move much slower and stick onto the dirt and dust in the sky when they run into them.

Would you ever imagine that giant sandstorms like these help make clouds and rain?

To get the air to cool, Hashem makes it rise. Air rises when it gets heated by the sun, when it runs into something big like a mountain, or when two large air masses run into each other and have nowhere else to go but up. The cooling of air as it rises is a process referred to as **ADIABATIC COOLING**. Once the air is cool enough, water molecules begin latching on to the cloud seeds, forming the water droplets that clouds are made of. These water droplets are big enough to reflect light, which is why we can see them as clouds, yet small enough that they don't fall to the ground.

If the conditions are right, the water droplets can grow much larger, fusing with other water droplets they run into.

Just think, little clouds like these may form right above your head when heat from the sun causes air to rise.

Here comes the rain train, with a shipment of water addressed to your door!

They get bigger and bigger until they are too heavy to stay aloft and fall as rain. Just imagine the amazing journey water may have taken. It rose from the ocean, then traveled thousands of miles before meeting up with a microscopic piece of a volcano that helped it become part of a great cloud before landing on your head!

We should all take some time to thank Hashem for the clouds in the sky! In addition to bringing us the water we need, they keep us cool by reflecting about twenty percent of the sun's rays. And as if that weren't enough, Hashem made clouds stunning so we can enjoy their beauty whenever we want. No need to go to a museum to enjoy artwork, just look up and you'll see a masterpiece made by Hashem!

CHOCHMAH V'CHESSED

The prophet Yeshayah instructed us to "raise up your eyes and see Who created all these" (*Yeshayah* 40:26). When we look up, we see giant mountains of water designed to transform from a gas to a liquid when they reach their destination! Is that for real? Could you ever in your wildest dreams imagine something like that was possible? Clouds and rain are absolute evidence of Hashem's ultimate power and unfathomable wisdom.

Hashem's art museum in the sky has free admission and new paintings every day!

CHAPTER 4

KOCHO U'GEVURASO MALEI OLAM

"Bereishis bara Elokim es ha'shamayim v'es ha'aretz" (Bereishis 1:1). There were beis (two) forms of fear (reishis is translated "to fear," as learned from the verse "reishis chochmah yiras Hashem") that Hashem made when He created our world. The first is Elokim, as we are scared to disobey Hashem's instructions lest He punish us. The second kind, which is the ideal and proper way to serve Hashem is "es ha'shamayim v'es ha'aretz," as when a person contemplates the great wisdom of Hashem all around us, he will come to an awe of Hashem and serve Him properly.

(Ohr HaChaim, Bereishis 1:1)

There are telephone poles lying in the street, uprooted trees everywhere, half the roof has flown away, and the basement has become an indoor pool! These are some of the telltale signs that a hurricane passed through your town. While hurricanes themselves may be scary, the way Hashem makes them is truly amazing.

Hurricanes form in tropical oceans when intensive heating from the sun causes lots of air and water to rise quickly, forming thick clouds and heavy rain. While this may be enough make a storm, only when wind speeds exceed 75 mph is it considered a hurricane. To understand how such strong winds are formed, we first need to know a little about **AIR PRESSURE**.

When Hurricane Sandy came to town in 2013, she wreaked havoc all over New York City.

Look how powerful the winds of a hurricane can be!

Even though the air around you seems empty, it is actually full of gases that are quite heavy. We don't notice the weight of air because we are so used to it, but it places about 15 pounds of pressure on every square inch of your body. When big storms form out at sea, much of the air near the surface of the water rises, so it's no longer putting pressure on the water below. This causes an area of very low air pressure. When this happens, air comes rushing in from the areas surrounding it to fill in the empty space. As long as the low-pressure region in the center of a hurricane remains, high-speed winds will continue to rush around it.

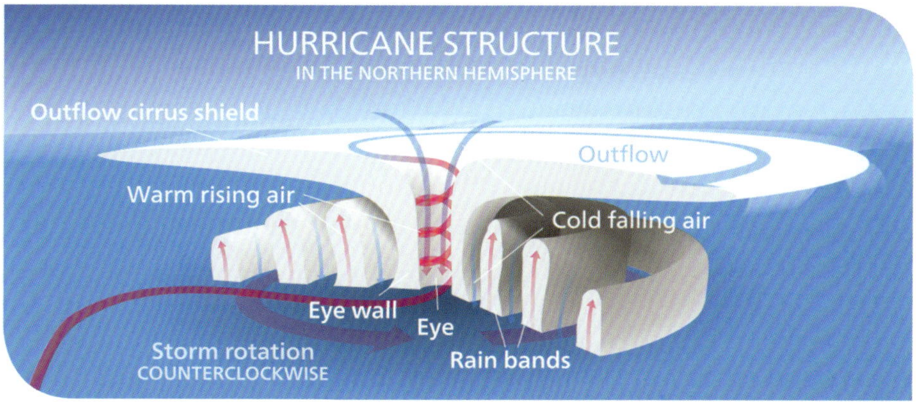

Heat of hurricane diagram

The center of a hurricane is called its eye because of the calm weather and lack of clouds in the middle of the storm, which make it look like an eye. While inside the eye may be calm, just outside is the **EYE WALL**, which has the strongest winds and is usually responsible for the most devastating effects of a hurricane. The eye wall gets its name because it literally looks like a wall of clouds surrounding the hurricane's eye.

This photo, taken from space after Hurricane Andrew hit Florida in 1992, gives you an idea of just how big hurricanes can be!

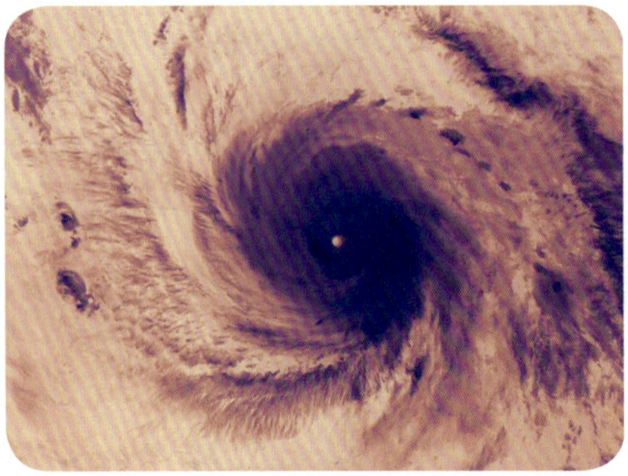

This is one eye you don't want looking at you!

Once a hurricane gets going, there is no stopping it—until it runs out of gas. The fuel driving the storm comes from warm water that evaporates and then condenses into clouds and rain, releasing heat into the air. This heat is used to lower the air pressure even more, causing faster winds and more rain. Hurricanes often get bigger and bigger out at sea until they hit land and have no more water to fuel their fury.

The amount of heat produced by a hurricane is enormous. It is estimated to be about two hundred times the entire world's electrical generating capacity or ten thousand nuclear bombs! Though the devastation of hurricanes is terrible, their heat is actually beneficial, as it helps maintain the warm and stable temperatures of our planet. Like smoke rising from a chimney, heat from a hurricane is released high up into the atmosphere through the top of the storm, where it is then distributed around the globe.

The world around us is full of marvels that Hashem made. Most of them benefit us in ways that are apparent, and when we think about them, they help us love Him even more. Others, like a hurricane, do more to put fear of Hashem into our hearts when we see how His power and might fill the world.

CHOCHMAH V'CHESSED

One way we see the Divine wisdom in giant storms is how they keep us humble. We find this in the Mishnah (*Kiddushin* 4:14) where it says that the majority of sailors are pious. The reason for this is that they are in constant danger from the dangers of storms at sea and are therefore humbled as they must look to Hashem for protection. I remember as a child watching the awesome power of Hurricane Andrew through a window; such experiences remind us of how frail we are. While the wind, clouds, and rain are generally signs of the blessings Hashem sends us, they are used when necessary to give us a "*potch*" and help keep us on the proper path.

CHAPTER 5

TORNADO ALLEY

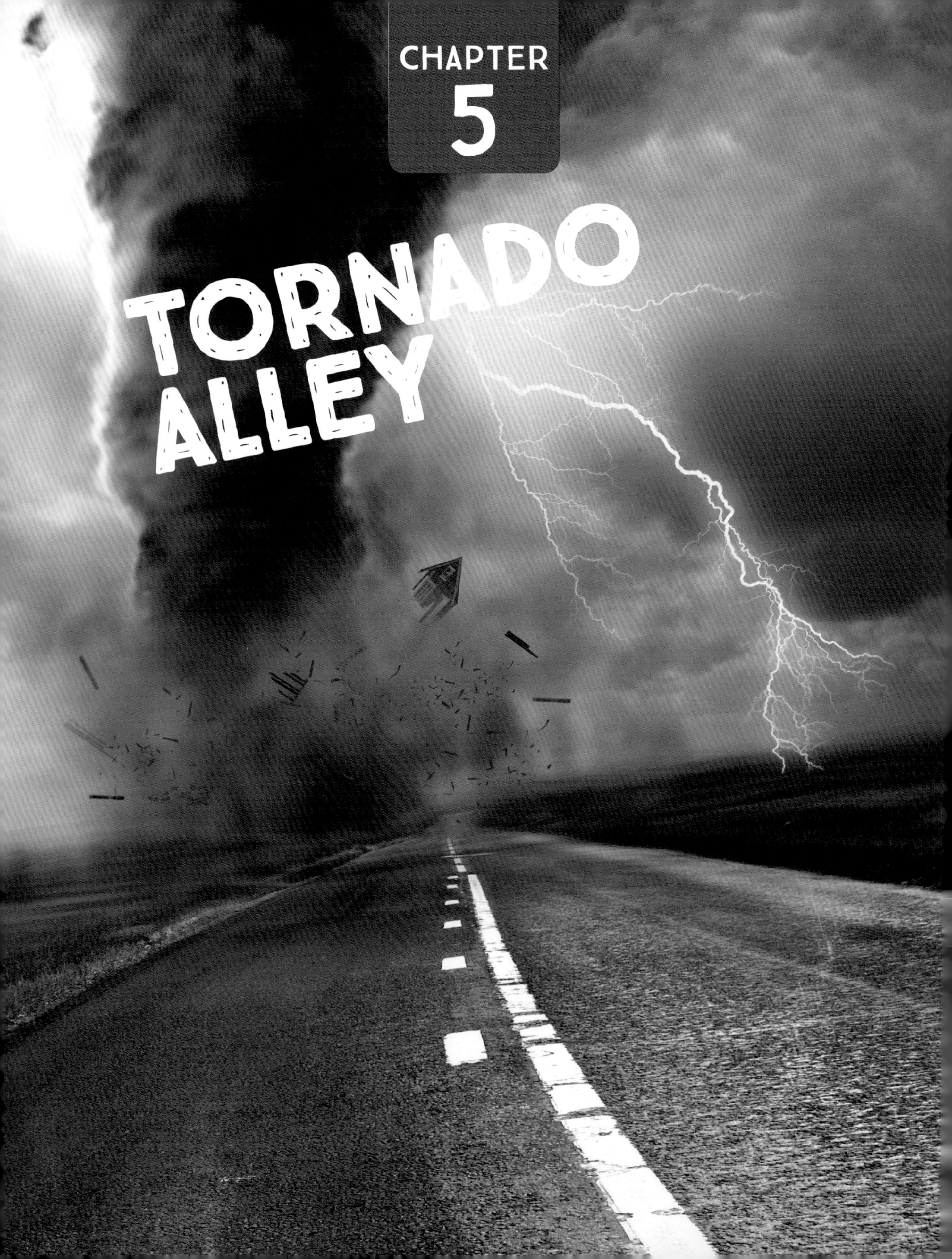

Ruach se'ara oseh devaro—The stormy wind fulfills His word.
(Tehillim 148:7)

Does the alley behind your house seem pretty boring? The occasional catfight or loose dog out looking for a snack are probably the only ones that liven up the little road. While most alleys are quiet and safe, there is one that is regularly visited by unwanted guests that cause a ruckus and destroy just about everything in their path. These terrors are **TORNADOES** and about a thousand of them come speeding through Tornado Alley every year.

Stovepipe tornado

Tornado Alley is the name given to the area of the United States that gets lots of tornadoes. The exact borders of Tornado Alley are somewhat disputed but everyone includes the states of Texas, Oklahoma, Arkansas, Missouri, and Iowa. If you're wondering why Tornado Alley gets so many tornadoes, you're in good company, as scientists have been trying to answer that question for a long time. While it seems to have something to do with the flatlands and hot climate of the states in Tornado Alley, scientists have been quite unsuccessful in their quest to understand how and why tornadoes form at all. Current attempts to solve the mystery of their mayhem include flying weather drones into them to try and find out what makes them twist. Perhaps in the near future, Hashem will reveal to them the secrets of tornadoes as He has with so many other of the marvels that fill our world. In the meantime, we will have to settle with some interesting facts we do know about them.

Tornadoes most often form within thunderstorms, where the air is being pushed and pulled in different directions. The stormy weather we often feel on the ground is nothing in comparison with what is going on inside the storm clouds themselves. Once in a while, a violently rotating column of air can form within the storm. Once it gets going, it takes on a cone shape and makes what we call a **FUNNEL CLOUD** emerging from the bottom of a stormy sky. If the funnel cloud keeps on growing until it reaches the ground, it earns the title of tornado. Once on the ground, tornadoes usually don't last for more than ten minutes—but with wind speeds that can reach over 300 mph, they can do an enormous amount of damage. The strongest tornadoes can tear homes off their foundations, topple buildings, and send cars flying a hundred yards or more.

Close encounters with tornadoes can be terrifying even for those in a safe place, as they can sound like a giant freight train passing, a great waterfall, or a tremendous roar outside. It is not just the sounds of tornadoes that vary, but the way they look as well. Their sizes range from a few feet to over two miles long on the ground and can come in all sorts of colors and shapes. They can turn red when formed over sandy

Funnel cloud

Wedge tornado

Dust-tube tornado

deserts, white when they fly over snow, black when they pick up a lot of dirt and debris, or blue when they form over water.

When a tornado forms over water, it is referred to as a **WATERSPOUT**, and there is actually reference to one in the Torah. We learn in *Sefer Yonah* that Yonah was on a ship that was caught in a great storm. As *Rashi* explains from *Pirkei d'Rabbi Eliezer*, the sailors knew it was a supernatural punishment because they saw the other boats sailing along

Waterspout

in peace nearby while they were stuck in a whirlwind. The *Malbim* explains that this was a tornado Hashem kept over Yonah's ship until the sailors tossed him into the sea. While all tornadoes are impressive, the one that wouldn't leave Yonah alone was probably the most awe-inspiring ever. The sailors saw for themselves that Yonah's G-d was truly the Ribbono Shel Olam, which is why they gave up their idols and became Jews.

CHOCHMAH V'CHESSED

The most fearsome force in nature was seemingly designed to impart fear. Unlike hurricanes, which do bring significant benefits to nature, tornadoes have no known ecological value. They serve as funnels of fury that remind us who's The Boss and keep us humble.

What's your favorite kind of tornado?

1. Stovepipe tornado
2. Dust-tube tornado
3. Funnel cloud
4. Waterspout
5. Wedge tornado

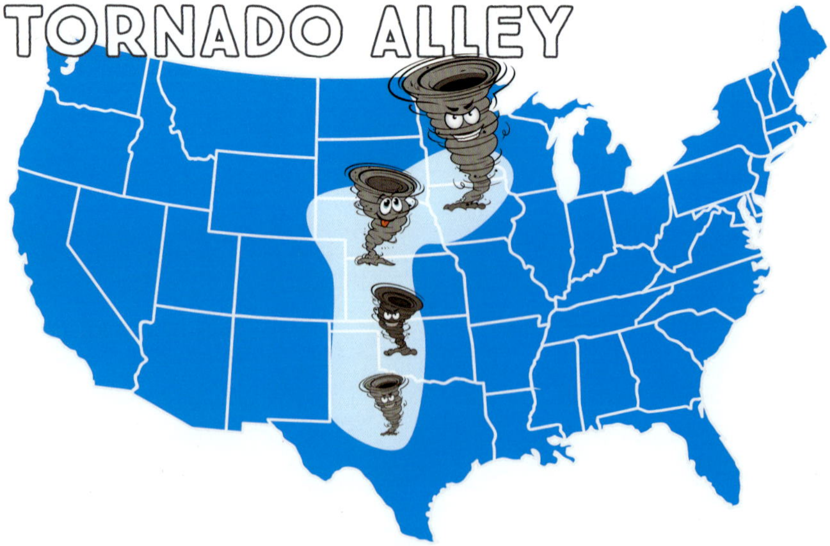

CHAPTER 6

FIRE IN THE SKY

Thunder was created only in order to straighten the crookedness of the heart.
(Berachos 59a)

Have you ever seen lightning light up the night sky and wonder what it was made of? Hashem's sound and light shows that accompany storms are truly amazing and many have shared your curiosity as to their cause. One such individual was Benjamin Franklin and he got the idea that lightning might be made up of electricity.

In 1752, he set out see if he was right by creating an experiment that consisted of tying an iron key to the end of a kite string. The key was then attached to a Leyden jar on the ground that could store static electricity (which is what we call electricity that isn't moving). If he was right and there was a huge amount of electricity in thunderclouds, his kite would pick up some of it when it flew through them, and send it down the wet string, filling the Leyden jar with electricity. When he saw that the jar had stored electricity, he knew he was right, and became famous for discovering what lightning was made of.

Ben Franklin's famous discovery of electricity using a kite and Leyden jar

If we want to understand how so much electricity gets collected inside thunderclouds, we need to know a little bit about the makeup of **ATOMS**, the building blocks for all matter in the universe. Atoms are collections of tiny bits of electricity called **PROTONS**, **NEUTRONS**, and **ELECTRONS**. Protons and neutrons are stuck to each other by way of a force Hashem made that holds them close together called the strong nuclear force. The protons and neutrons don't move around much and they make up the nucleus or center of the atom. Electrons are found at a distance from them and are always busy, whizzing in orbits around the nucleus at incredibly fast speeds. The reason why they don't fly away is because of their electrical attraction to protons. Protons have a positive electric charge, similar to the positive end of a magnet, while electrons have a negative electric charge, similar to the negative side of a magnet. Just like the positive end of one magnet is attracted to the negative side of another one, electrons are attracted to protons, keeping them in their orbits. While this attraction is pretty strong, if something runs into the electrons hard enough, they can get knocked out of orbit and fly away.

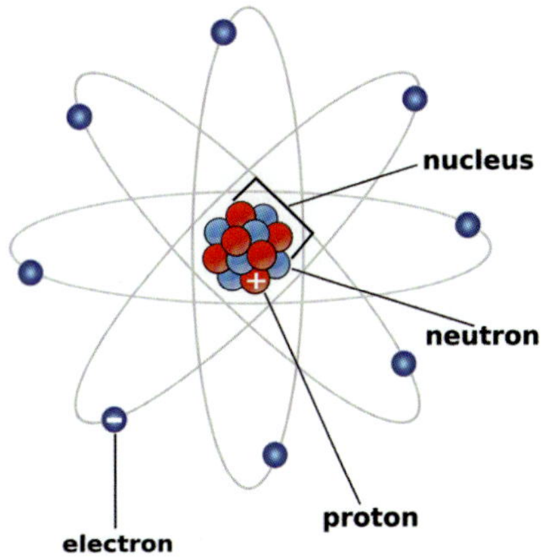

Amazing atoms fill our world and often lend their electrons to help make lightning.

Within thunderclouds, ice crystals often collide with water droplets. These collisions knock many electrons away from their atoms. The electrons, which are little bits of electricity, tend to stick to the larger water droplets in clouds, filling them with electric charge. Because the water droplets are heavier than the ice crystals, they tend to concentrate at the bottom of the clouds, while the light ice crystals

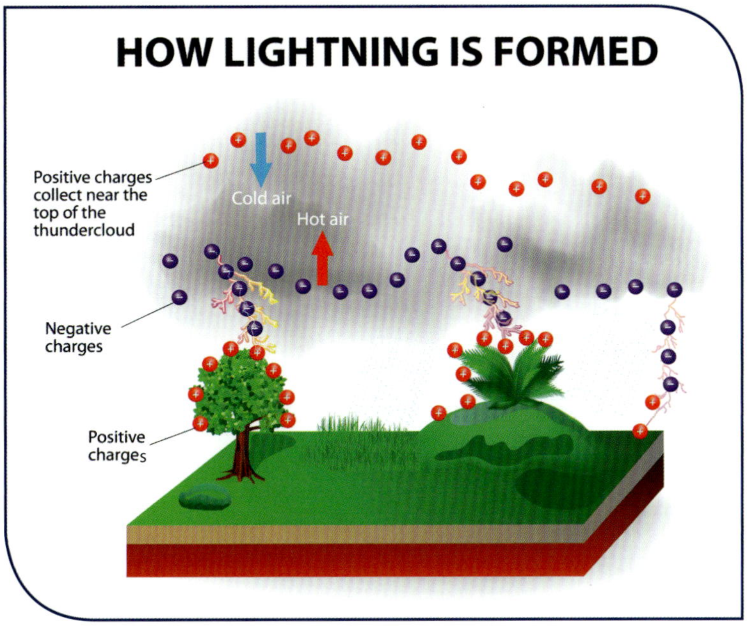

It's the buildup of electrical charges that lights up the sky.

concentrate at the top of the clouds. As more and more electrons pile up near the bottom of the cloud, the electric charge becomes so great that it can no longer be held within the cloud and spills out in the form of lightning.

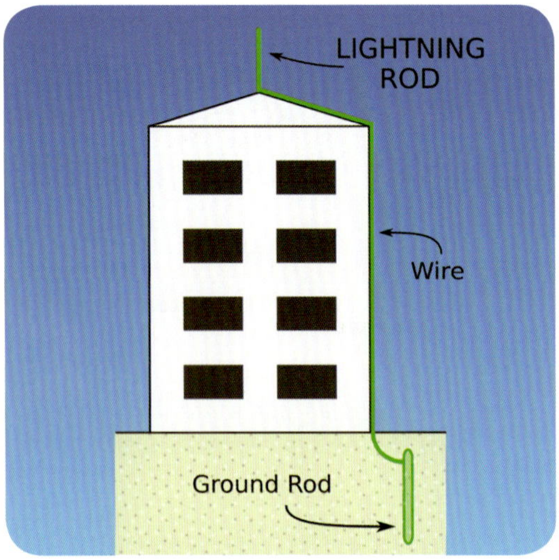

Metal lightning rods, invented by Ben Franklin, have saved more buildings from burning down than firefighters have by allowing the electricity of a strike to pass safely through them into the ground.

Ben Franklin's understanding of the nature of lightning and electricity was quite impressive and he even knew that it was a form of fire, unlike his contemporaries who viewed it as something different. There are many Torah sources that suggest that he was right. One example is *Bamidbar Rabbah* (15:17) that states: "Lightning is a product of the heavenly fire which shoots out and lights up the world." Electricity is truly a great gift from Hashem. This becomes apparent if we take some time to think of all the ways we benefit from it, such as lighting up our homes and powering our air conditioners. Like with any present we get, it is appropriate to say thank you—so don't forget to thank Hashem for making electricity a part of our world and allowing us to understand and harness its power.

CHOCHMAH V'CHESSED

I still remember as a young child watching the rain out of my friend's living room window when, bang! A bolt of lightning struck right before our eyes. It hit a tree in his yard and blew a lot of tree bark into his pool. We both fell off the sofa from the blast, which was a truly awesome sight that I will never forget.

Encounters such as these with Hashem's power are great opportunities to wake us up to the reality of His presence and strength. This is why the blessings "*Oseh maaseh bereishis*," which we say when we see lightning, or "*Kocho u'gevuraso malei olam*," which we say when we hear thunder, were instituted by our Sages so we could utilize these precious moments to connect to and praise Him. It seems to me that since thunder and lightning are far less dangerous than other displays of G-dly power (such as hurricanes, tornadoes, and earthquakes), they were chosen to be the most regular reminders of the Creator's presence and power.

Isn't it incredible how Hashem fills clouds with electricity that light up the sky!

While lightning can cause damage, it also benefits our planet in a number of ways. **NITROGEN** is a chemical that all life needs to live and is one of the key components of our DNA. Even though most of the air around us is made up of nitrogen gas, it is in a form that is unusable for most forms of life. Lightning heats up the air around it to temperatures of 50,000 degrees Fahrenheit,

which fuses nitrogen atoms together with other atoms, and, through a process called nitrogen fixation, transforms it into a form that can be used.

Forest fires are primarily caused by lightning and may seem pretty awful, but occasional ones are actually very beneficial to the environment. In addition to removing dead and decaying plants that act as fuel for the fire, the intense heat strips the nutrients and minerals out of burning plants and puts them into the ground, making rich soil for new species to grow. We see how these fires are clearly part of the Divine design as there are certain seeds that lie dormant for many years and only grow after a big blaze.

Forest fires can be very scary, but even they have their role to play in Hashem's world.

CHAPTER 7

REMARKABLE RIVERS

When you see mountains, valleys, or rivers, you recite the blessing praising He "Who made the act of Creation."
(Berachos 54a)

Have you seen a river lately? If you did, I hope you thanked Hashem for these long winding treasures of goodness that are scattered throughout our world. If you never realized how great rivers are, keep reading and prepare to be amazed!

Do you like going fishing? One of the best places to fish is in rivers because they are often teeming with fish (and other living things), making them an important source of food for the people and animals of our planet. Perhaps even more important than the food they provide, rivers bring fresh water to countless creatures around the globe.

Did you ever try carrying a tree log? If you did, it probably wasn't fun as logs are extremely heavy and difficult to move. Despite the burden of their transport, people have always needed a lot of wood for things like building homes and making furniture. How did they move trees before trucks? Often, they used rivers. Lumberjacks would throw the trees they cut down into the water and then drive them downstream to lumber camps where the wood was processed. While log driving has mostly been replaced by trucks and trains, rivers still play a significant role in the transport of people and goods today.

Log driving was once the quickest and least expensive way to transport trees.

Did standing on your driveway ever remind you of a river? It is quite possible that the gravel that was used to make your driveway came from the bottom of a river. It's not just driveways that are made from river products; the gravel and sand formed in rivers are used in many areas of construction. Rivers are so packed with beneficial things that even the dirt at their bottom is useful!

What about sitting in your cool air-conditioned house on a hot summer day; does that remind you of a river? Well, if you are living in New York, or any of the other thirty-four US states that have **HYDROELECTRIC POWER PLANTS**, it is very likely that much of the energy powering your home comes from a river. In New York, the Robert Moses Niagara Hydroelectric Power Station in Lewiston converts the power of flowing water in the Niagara river into electricity, providing a significant portion of New York's energy needs.

Isn't it incredible how hydroelectric power plants can use rivers to light up your living room!

Have you ever been white-water rafting? If not, you may want to try it, as it's a lot of fun. In addition to all the practical benefits we enjoy from rivers, they help make many people happy as they go swimming in them—or go rushing down them on rafts.

Rivers can be a whole lot of fun!

While they can be thousands of miles long, rivers eventually come to an end — in what is referred to as the **MOUTH** of the river — where they pour into oceans or lakes. This is really important because if not for rivers, rainwater would have no way of getting back to the oceans and lakes from where it came and it would flood the ground. Rivers function to drain the land of excess water in addition to the abundance of blessings they bring.

Covering forty percent of South America, the amazing Amazon is the largest river in the world.

It is truly amazing how many aspects of our lives are affected by rivers. If we take some time to reflect upon them, we will surely see signs of Hashem's great wisdom and kindness. So the next time you pass by a river, don't pass up the opportunity to thank Hashem for the remarkable rivers that He made.

CHOCHMAH V'CHESSED

To appreciate the benefits of rivers a little bit more, we can reflect upon the importance of proper drainage in cities and the unpleasant consequences that occur when it doesn't work efficiently. In cities after large storms, we see how gutters can get backed up, making canoes the easiest ways to travel. Land, much like your home or city, would be flooded if not for proper drainage that rivers provide and this could arguably be their most important function.

Modern advancements in plumbing and drainage have had a huge impact on the development of civilization and are surely worthy of praise to those who developed them. What is far more impressive, however, is the drainage system Hashem designed for our world. In addition to rivers' superb performance of draining excess water so that land stays solid, they bring a host of benefits to us, such as the ones listed above. To top off their display of Hashem's wisdom and kindness, when they finally end and meet the sea, they form a special habitat called an **ESTUARY** that has great ecological importance. The combination of fresh and salt water, along with the partial flooding of the area, create an ideal environment for many birds, fish, and other creatures to lay their eggs and raise their young.

Sometimes it takes a disaster to appreciate simple things like having ground to walk on.

Estuaries like this one help many creatures find the perfect home.

CHAPTER 8

MOUNTAINS OF ICE

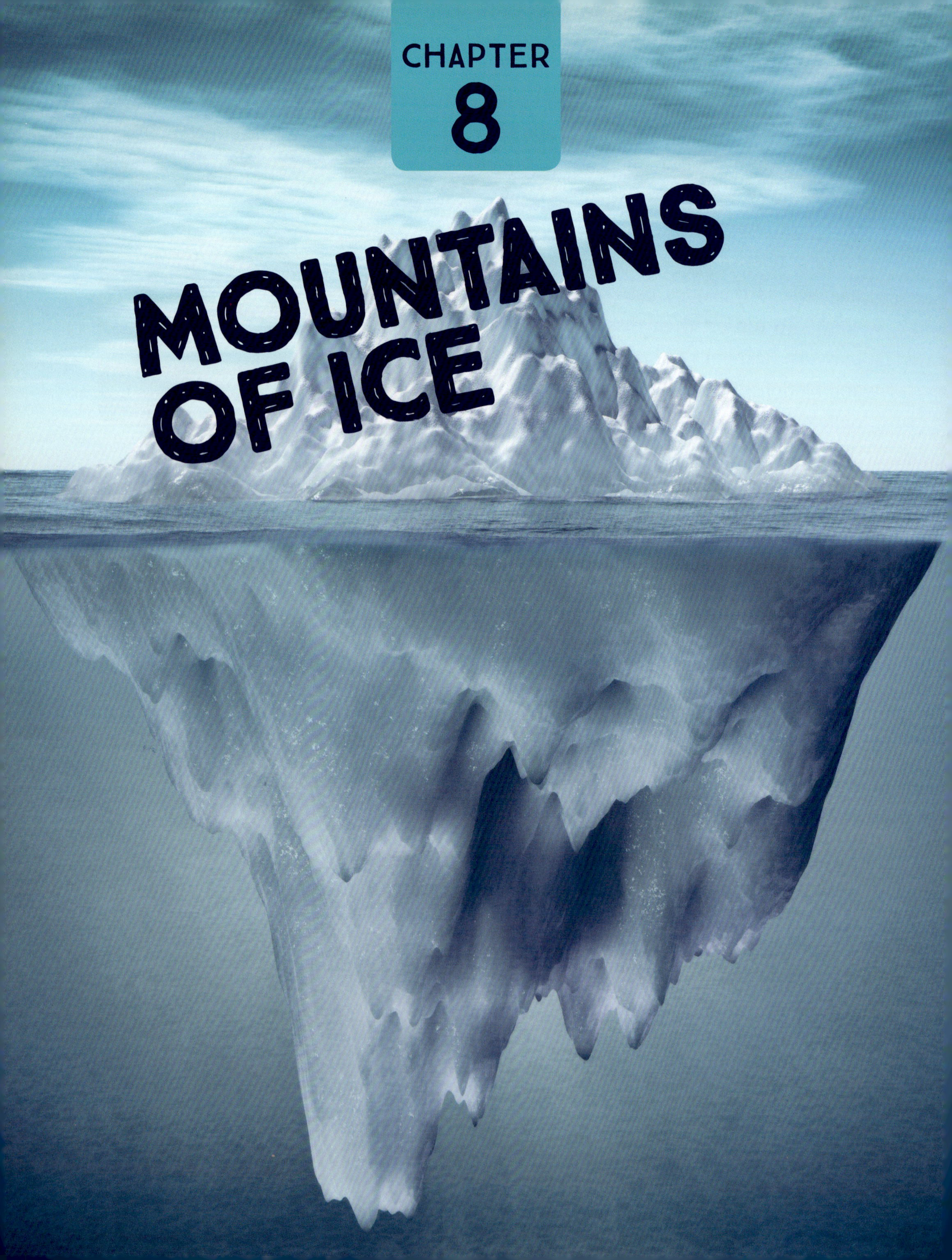

Hashem made the world for the purpose of teaching us about Him and it was His wish that we should study the world and see His greatness.

Rabbi Avigdor Miller (Ohr Avigdor, Shaar HaBechinah, Chovos HaLevavos)

Sometimes, you just have to marvel at how Hashem thought of everything when He made our world. Think about how much water is around during the winter when it rains. All the plants and animals have fresh water in abundance during this time, but what happens in the summer when the heat of the sun has dried up all the rain, and the rivers and lakes that were once overflowing begin to dwindle?

Dried-out riverbed

To ensure His creatures get the fresh water they need, Hashem stores huge amounts of fresh water every year in His "big freezer"—in the form of snow and glaciers. **GLACIERS** are solid blocks of ice that form in places where not all of the snow melts in the summer. Each year, more and more snow piles up and its weight pushes down on the snow underneath it, turning it into solid ice. In the summer,

In addition to storing water, snow-covered mountains can be a lot of fun!

Hashem defrosts His freezer a bit and the melted water from snow and glaciers flows into and replenishes many of the important rivers and lakes of our world.

In addition to filling up rivers, glaciers are among the most impressive features of our planet. They are huge! The biggest ones can be over a thousand miles long and a mile thick. Despite their monstrous size, glaciers are constantly on the move, often from sliding over a thin film of water that forms at their bottom. While they aren't known for their remarkable speed, usually a few hundred meters per year, the way they shape the land they pass over is extraordinary. Because of their enormous size and weight, they can totally change the features of the land they travel over by grinding and crushing rocks of all shapes and sizes. It happens sometimes that big chunks of glaciers fall off into the ocean. They look like floating mountains of ice, which is why they were named **ICEBERGS**, meaning "ice mountain" in Dutch (*ijsberg*).

Icebergs can be hundreds of miles long and over two thousand feet tall but only a small fraction of that is seen above the water. Because of the time they spent on land

This is literally just the tip of the iceberg!

A polar bear hunting out on the ice

tearing apart and picking up rocks, icebergs are able to help microscopic plants and algae by providing them with the iron they need to grow.

Just like we need basic vitamins and minerals, so do the creatures of the sea. Most living creatures get their mineral requirements from drinking water that flows over rocks and picks up minerals like iron contained within them. When rivers flow into the oceans, minerals that are necessary for living are brought to creatures that live in the sea. Near the North and South Poles, however, there are no rivers with flowing water because everything is frozen. To replace them, Hashem made giant vitamins, the size of mountains, which melt as they sail their way around the icy waters, providing the sea life there with its nutritional needs.

Icebergs should not be confused with **SEA ICE**, which forms in the open ocean near the North and South Poles when the surface water freezes over. In addition to providing a home and hunting ground for animals like polar bears, seals, walruses, and penguins, sea ice is also used to help sea creatures in an amazing way!

When sea ice forms, the salt from the seawater doesn't freeze with the ice and falls out underneath it. The increased salt makes the water heavier and it sinks to the bottom of the ocean. When the sinking water hits the bottom of the sea, it replaces the water that was previously there and forces it up to the surface. This process brings nutrients that were at the bottom of the ocean up to the sea surface where it helps the ocean's food chain flourish. So the next time you eat fish, in addition to thanking Hashem for the yummy food, you can remember how He provides for the needs of every living thing in truly wondrous ways.

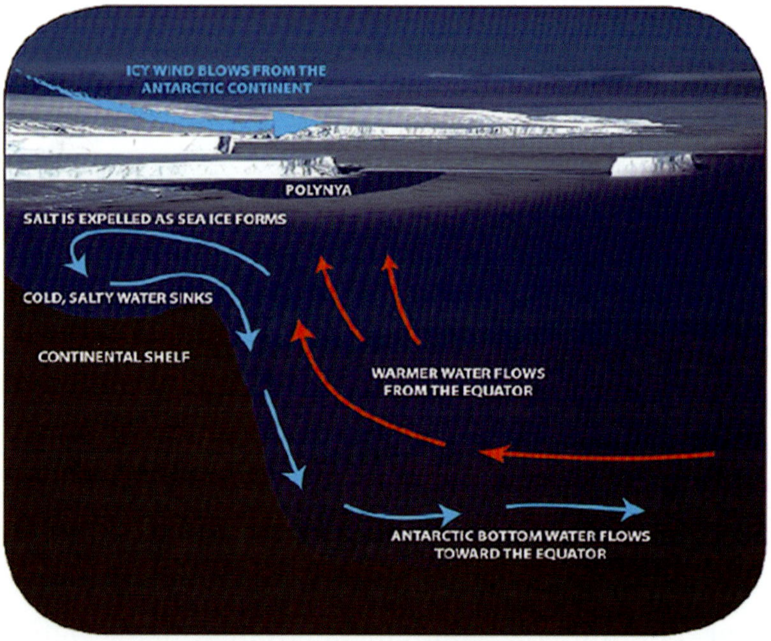

Hashem's Super Power Pump is strong enough to push water around the entire world, has no electric bill, and runs on salt!

CHOCHMAH V'CHESSED

"You open Your hand and satisfy every living thing [with] its desire" (*Tehillim* 145:16). A great example of what David HaMelech was talking about is how Hashem pushes giant icebergs out to sea, sending the tiny plants which are the base of the ocean's food chain, called **PHYTOPLANKTON**, the iron they need to grow. In general, rivers bring to the ocean the nutrients needed to support the growth of phytoplankton. In the frozen continent of Antarctica, there are no rivers to carry out this vital task. Scientists have recently found that an icy equivalent to a nutrient-bearing river exists, which they have termed "iceberg alley." Icebergs are an ingenious alternative to rivers to get phytoplankton the nutrients they need!

While icebergs help support life in the sea, without sea ice there might not be much life in the ocean at all. By designing water to become heavier as it increases its salinity (salt content), the cold and salty water that is formed along with the sea ice puts motion in the ocean. This is necessary because sediment at the bottom of the ocean contains the nutrients from plant and animal matter that died and decayed. If not for the sea ice, those nutrients would be lost forever. Again, we see unbelievable wisdom, enabling a massive amount of water transport with such a simple design.

As far as kindness goes, I don't know about you, but I really like gefilte fish and salmon, which may not be around if not for the great mountains of ice. All life in the sea is testimony to how great a present frozen water can be.

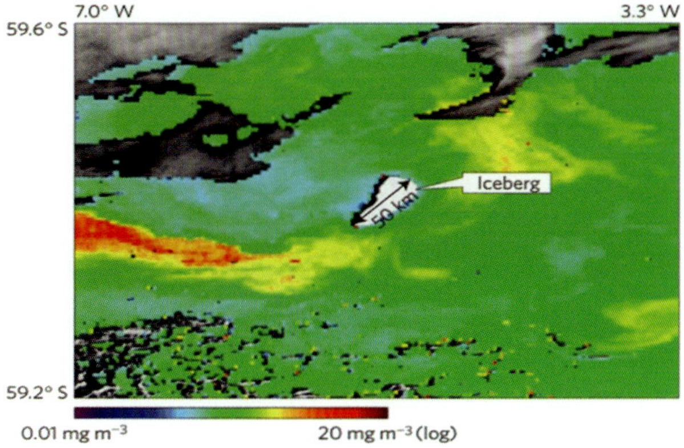

Figure 1 | Chlorophyll-*a* concentration on 12 January 2013, from the MODIS Aqua satellite. Giant iceberg C16 is visible in the centre of the picture, with enhanced levels spreading southwest and northeast from the iceberg. Greyscale areas show cloud cover.

This infrared image shows red and yellow colors, which are signs of photosynthesis taking place, illustrating how icebergs cause life to flourish.

CHAPTER 9

SENSATIONAL STORAGE

Through the wise man's study of creation...he will recognize God's constant goodness, and these insights serve as a source of life, for they bring him to fear, love, and serve God...

Rabbeinu Yonah (Derashos, Parashas Tazria)

We have already learned about many of the freshwater storage systems Hashem built for our world like rivers, lakes, glaciers, and snow. There is another one that is perhaps the most sensational of them all—and is right below our feet!

All you need is a shovel and some perseverance to dig your own well.

If you want to see where Hashem keeps most of the water that is used to support life on our planet, just take a shovel and start digging. Eventually you will see there is a lot of water beneath the earth's surface. It may be hard to imagine, but there is actually a hundred times more fresh water underneath the ground than in all of the rivers and lakes of the world combined. We call this water **GROUNDWATER** because it is located under the ground and is found almost everywhere; beneath hills, mountains, plains, and even deserts.

You may not have a pool in your backyard but you probably have a big one beneath it.

Why do you think Hashem put all that water underneath the ground? One of the reasons is surely so that trees and plants will have the water they need. Most vegetation grows far away from rivers and lakes, so if not for groundwater they would have a hard time surviving—and making food for us to eat. To give you an idea just how much water plants need to grow, it takes about 150 gallons of water to produce enough wheat to make just one loaf of bread. Because the ground has so much water buried beneath it, plants are able to grow just about anywhere.

It isn't just the trees that benefit, as the water you and your family use on a regular basis probably comes from groundwater. Most of the water used in homes, called tap water, comes from groundwater as it is usually much cleaner than rivers and lakes that are often polluted. In addition, it provides water for people living in climates that don't have large lakes and rivers to supply them with water. Other uses of groundwater include irrigation, mining, and industrial uses.

How can the ground hold so much water? Hashem accomplished this fantastic feat by making the soil and rocks underneath the ground with cracks and spaces that are able to hold water. When enough water exists that it can be used by people by digging a well, it is called an **AQUIFER**. Aquifers get refilled after it rains when the water seeps into the dirt and rocks below the surface. It keeps sinking until it reaches what we call the **WATER TABLE**, which is the point in which the ground becomes saturated (flooded) with water. Hashem made the water table far enough underneath us that the ground we walk on

Water wells like this one enable people to live just about anywhere.

would be dry and hard, yet close enough to the surface that it is accessible when we need it. Amazing! Once rainwater reaches the water table it usually doesn't stay put, but creeps along until it makes its way to a well, natural spring, or river that carries it back toward the ocean where it came from.

In the Torah, the gathering of water below the earth is referred to as the *tehom*. We see how important this water is because the greatest *simchah* in the Beis Hamikdash was on Sukkos, when water was drawn from the Shiloach spring and brought to be poured as a special libation on the *Mizbei'ach*, the *nisuch ha'mayim*. The reason the *nisuch ha'mayim* was done on Sukkos is because the world is judged regarding water on Sukkos. Hashem gave us a mitzvah of pouring water that came from the *tehom* on the *mizbei'ach* so that the waters of the *tehom* and rainwater would be a *berachah* for us in the year to come.

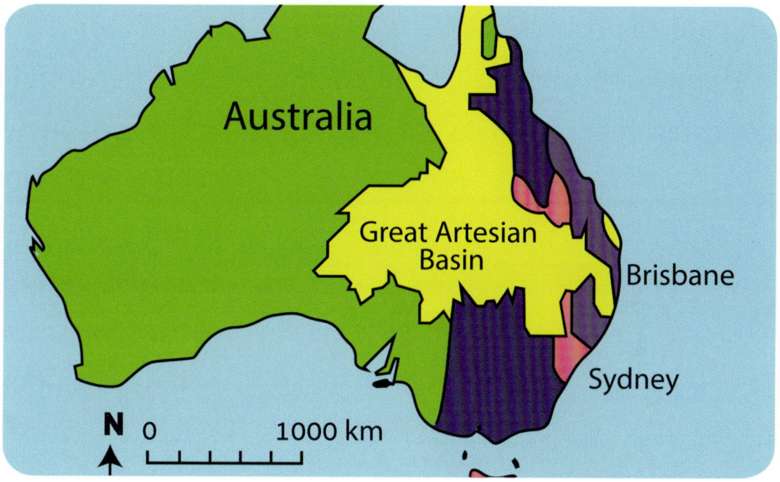

At 660,000 square miles, the Great Artesian Basin is the largest aquifer in the world and provides much of Australia's freshwater needs.

Water from a natural spring like this one tastes great and is full of minerals that were brought up from underneath the earth.

CHOCHMAH V'CHESSED

Imagine you were an engineer. You were given the task of building water-storage tanks for every city on earth. Now, it is not in your budget to regularly maintain these tanks, so they must be made in a way that they will always have water, never break, and are easily accessible.

If a person were to invent such a system, he would receive worldwide praise and get every medal and award around. In fact, a special award may even be made up for such a feat. Does Hashem deserve less? The water table that is buried beneath you, and I mean literally wherever you are right now, is testimony to an awesome Divine design. The soil is a superb water-storage device that holds twenty trillion gallons of fresh water for the needs of plants, animals, and people, never breaks, is deep enough to avoid the heat of the sun that would dry it up, yet close enough to access, and it actually filters the water as it moves through the soil.

As far as kindness goes, most of the grains, fruits, flowers, and trees of our world are gifts that depend on groundwater. For a more direct benefit, the great taste and nutritional value of natural spring water comes straight from groundwater. After picking up important minerals like iron, zinc, manganese, copper, fluoride, iodine, chromium, and selenium, spring water comes out—ready to drink and enjoy!

CHAPTER 10

RING OF FIRE

The beginning of wisdom is the fear of G-d.
(Tehillim 111:10)

Did you know that the ground beneath you is floating around like a boat? We don't feel it as it's going pretty slow, about three inches a year, but the continents we live on sail over a fiery sea of melted rock we call **MAGMA**. Magma forms deep beneath the earth's surface where rock melts due to the high temperatures and pressure down there. While we can't see magma, we know it is there, as it occasionally comes bursting out of volcanoes.

What type of treasures do you think this volcano is sending our way?

Is it hard to believe that the continents sail around the world on a red-hot sea? If you need some convincing, take a look at a world map and you will notice how the continents seem to fit together like pieces of a puzzle. Scientists were aware that the continents seemed to have drifted apart, but for a long time they had no idea how they could have moved. In the 1960s, convincing evidence showed the continents were moving and the theory of **PLATE TECTONICS** was proposed, suggesting that currents of boiling magma beneath the earth's surface caused the land masses that the continents rest upon to drift apart.

Are you a bit skeptical about scientific theories? If you are, I commend you! Scientists change their minds a lot and make many assumptions that may not be true. However, with regard to the continents sailing apart over a molten sea, it seems that they got this one right. The *Zohar* (12:1) teaches us: "One single continent came out of the water, and from it seven continents were formed." Similarly, we learn how the surface of the earth is divided into floating plates in *Pirkei d'Rabbi Eliezer* (chapter 5): "The earth is made up of plates covering the depths, as a ship that floats on the sea."* While we can commend the scientists for their discovery, they could have saved a lot of time and money by just asking some rabbis to help them solve their puzzle!

Once you get the concept that the earth's surface is made up of enormous plate-like sections of rock that move in different directions, you can understand other features

Here we see some of the damage caused by a giant earthquake in Thailand.

* Taken from *Mysteries of Creation* by Rabbi David Brown.

of our world, like **EARTHQUAKES** and **VOLCANOES**. Earthquakes happen when two plates slide past or collide with one another. The force is so great that the ground between the two plates breaks, forming what we call a **FAULT**. Tremendous damage results from the ground splitting and can cause roads and bridges to break, buildings to collapse, landslides, and tsunamis.

Similarly, the explosive eruptions of volcanoes are most often caused by two plates that run into each other. One of the plates gets pushed far underneath the ground where the rock melts into magma. The hot magma then rises to the surface where it flows out of a volcano. While it is really the same stuff, when magma comes up to the surface it is called **LAVA**—and can cause great damage by melting just about everything in its path.

Most active volcanoes are found along the edges of the Pacific Ocean where the rock that the ocean rests on runs into neighboring plates. A string of 452 volcanoes

surrounds the ocean and is referred to as the Ring of Fire. While the Ring of Fire may be far away from you, the products of volcanoes are probably not. Lava is the source of many important resources like building materials, diamond, gold, silver, copper, aluminum, and microscopic glass shards that are put in some toothpastes to help clean teeth. So the next time you see cinder blocks, gold, silver, or you are just brushing your teeth, you can think about volcanoes and the marvelous ways Hashem moves mountains to provide us with our needs.

Measuring 25,000 miles long, this is surely the biggest ring on earth.

CHOCHMAH V'CHESSED

If you have ever felt the earth rumble, you know how earthquakes can help you do *teshuvah* really fast. Just in case the ground shaking didn't shake sinners up enough to listen to Hashem, the fiery sea of magma beneath the earth's surface is waiting to help purify them of their sins. Volcanoes and other glimpses we get of the intense heat below us, like hot springs, serve as regular reminders of how wise it is to stay on the proper path.

While the reminder of the place where the wicked will be cleansed from their sins is apparent when we experience earthquakes and witness volcanoes, the host of gifts that come

There are all kinds of treasures brought to us by volcanoes.

Lava lakes like this one are pretty cool, but way too hot to swim in!

from the fires below are often not. Much of the treasures we enjoy like gold, silver, and diamonds are brought up from the depths for us to enjoy by way of volcanic eruptions. This is coupled with the fact that volcanic ash takes part in forming clouds and rain, which helps remind us that the One who sometimes needs to punish us, even in the depths of Gehinnom, is doing it out of true kindness and love.

CHAPTER 11

HIDDEN TREASURES

[It is a mitzvah] to contemplate the wonders of His creations in order that they should be a sign to remind us of Him.

Rabbeinu Bachya (Introduction to Chovos HaLevavos)

Are you up for an adventure? How about a hunt for hidden treasures? Let's go together to find out what riches Hashem buried beneath us!

The first stop on our quest is the Appalachian Mountains. Black lines of coal that look like veins are visible in the mountains all around as the region has one of the largest sources of coal in the world. You may not think of coal as a treasure, but it is often referred to as "black diamonds" because of its great value. Coal burns well, as you may remember from your last barbeque, and its heat is used to produce over 40 percent of the world's electricity, make steel, and manufacture cement. While coal mining is lucrative, you need to be careful underground, as the roofs of coal mines have been known to cave in unexpectedly. In addition, methane gas is lurking around, which can cause sudden explosions. No camp fires here!

Continuous mining machines are amazing as they can extract five tons of coal in one minute.

Our next destination is Carlin, Nevada. The rocks around here may seem ordinary to you, but any gold prospector will tell you they are worth a fortune. Carlin is home to the third-largest gold mine in the world, having produced close to five thousand tons of gold since it was discovered. The reason why it is hard to identify gold is that it is most often found in small quantities within rocks we call **ORES**. In fact, most precious metals are found within rocks and have to be melted or treated with chemicals to get them out.

Coal miners used to take canaries with them into mines as they are super-sensitive to methane gas and if they died, the miners knew it was time to get out.

If the rocks in your backyard look like this one, you could literally be sitting on a gold mine.

While we are in Nevada, we can pick up some silver, as there is plenty to be found. There is so much silver in Nevada it is nicknamed the "Silver State" and it's quite possible the silver in your Kiddush cup or Shabbos candles came from one of the mines here. In addition to gold and silver, Nevada mines are rich in copper and iron. Some uses of copper are water pipes and the wires that conduct electricity in your home and appliances. Iron is used to make a host of things like cars, buildings, benches, and pans.

This trip wouldn't be complete if we didn't visit Idaho, which is known as the "Gem State" because nearly every type of gemstone has been found there. Gems are pretty

With temperatures reaching 2,000°F, smelting pots turn ores into fiery soups of liquid metal and rock to help separate them.

easy to find if you know where to look, and all you need is a shovel, hammer, and a lot of patience sifting through the sand. Be careful which rocks you throw out, as it is hard to recognize a red ruby or blue sapphire when you see one. They have to be professionally cut and polished before one can really appreciate their beauty.

While our expedition underground together has come to an end, your personal journey to see the kindness of Hashem around you has just begun. Try noticing things around your house and outside that were made using metals or rocks that were mined. You will soon realize that the city you live in is full of things that are only possible because Hashem put so many useful materials for us in the ground. Metals found in rocks weren't put there for plants and animals because they have no use for them. There were put there by Hashem just for us. He knew that we would need them to make cars, buildings, planes, trains, computers, phones, and more, so He put them underneath the ground where they would be available for us when we were ready to use them.

CHOCHMAH V'CHESSED

If you want some practice contemplating the world around you to see signs of Hashem's kindness everywhere, try the following exercise; it amazes me every time. As you are walking down a sidewalk, think about the bricks beneath your feet. These great gifts came from the treasures hidden in the earth that enable us to have smooth roads to walk on. If you are in a city, look in any direction and you will see houses or buildings that tower above the earth thanks to the metals like iron, which are used to reinforce the structure. While iron is not usually thought of as a "precious metal," it is the main ingredient used to make steel, which most buildings rely upon to stay sturdy. In Israel, "Jerusalem Stone" decorates the outside of many homes. Stone comes straight from Hashem, as these little pieces of mountains just need to be chiseled a little to make buildings beautiful.

At some point in this excursion, you will probably walk into a house or building where you might notice some really fancy floors. Floors are often made from ceramic, granite, or marble, all of which come from the ground. Enter the kitchen, and the granite counters, formed from cooled-off magma that once flowed from

What is your favorite gift from underneath the ground?

a volcano, may both impress you with the beauty of Hashem's artwork as well as the wisdom that went in to making this substance so easy to clean.

The most precious stones on earth

This exercise is a lot of fun. When we begin looking around for all the places we benefit from rocks, metals, or cooled-off volcanoes, we'll start to realize how we are truly surrounded by signs of Hashem.

CHAPTER 12

GREEN GIANTS

Then shall all the forest trees shout for joy at the presence of the Eternal One, who is coming to rule the earth; God will rule the world justly and its people in faithfulness.

(Tehillim 96:12–13)

Have you ever had the pleasure of playing with a giant? It can be lots of fun! What kind of giant?

Green giants are easy to find because trees grow just about everywhere and they let you climb them, swing on their branches, and sit in their shade.

Now that's a real giant! Giant sequoia trees are extermely tall, and the one shown here is the largest tree on earth.

You may not think of trees as being alive because they don't move around, but they are really quite active living things. They don't need to run around chasing after meals because of a great gift Hashem gave them: the ability to make their own food! With a little bit of sunshine, air, and water, trees are able to make sugars that give them the energy they need to live and grow. We call a tree's ability to use the energy of the sun to make food **PHOTOSYNTHESIS**, and this is one of Hashem's many brilliant inventions for our world as it enables trees and plants to grow just about anywhere.

Have you ever wondered why leaves are green? White light coming from the sun actually contains many different colors, as can be seen sometimes in a rainbow. The sunlight-catching devices within

What is your favorite way to have fun with trees?

leaves are called **CHLOROPLASTS**, which can absorb most of the colors of light and use it to power the microscopic sugar production within leaves. They can't, however, absorb or use the green light coming from the sun. Instead of going into the leaves, the green light bounces off of them and enters your eyes, making leaves look green.

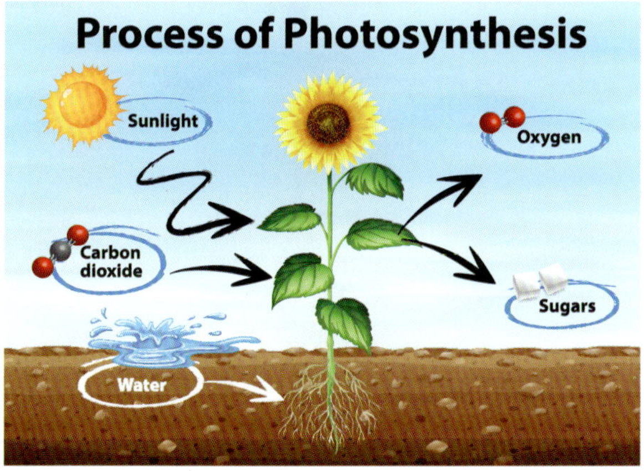

Sunlight and air are pretty easy for trees to find. Water can be much more of a challenge. This is especially true for large trees that can use up to two hundred gallons a day! To enable trees to drink all that water, Hashem gave them roots with tiny hairs at their ends that soak up water and nutrients from the ground. Since some places get lots of rain and others get very little, Hashem made two kinds of root systems to ensure trees would grow all over His world:

- Trees with **TAP ROOTS** grow downward very far beneath the ground. They do well in places that don't always get enough rain, as they can reach reservoirs of water far below the surface.
- Other trees have lots of shallow roots that grow outward from the tree called **FIBROUS ROOTS**. They are better suited for places that have an abundance of rain as they have enough water in the ground near the surface and are able to soak up huge amounts of water with all of their roots.

Did you ever imagine how much of a tree was underground?

Trees are truly amazing creations. Because there are so many of them, we have countless opportunities to praise Hashem for His wondrous works. Perhaps the best place to appreciate trees is within your home. Look around and you will see so many things that come from trees, like bookshelves, chairs, tables, beds, cabinets, apples, peaches, pears, and more. Even the paper you are reading right now came from a tree! Seeing that Hashem was so generous in making trees for us, the least we can do is say thank you once in a while for making these great green giants a part of our world.

CHOCHMAH V'CHESSED

It is hard to rate which of Hashem's creations is the most impressive, but making a being that eats air and sunlight has got to be up there on the list. Let's not forget about the fact that trees can live thousands of years and grow over three hundred feet tall. The Divine wisdom we see in plants is awesome and their Creator is certainly worthy of awe!

When it comes to kindness, they are no less impressive; just thinking for a couple of minutes how much we benefit from plants will boggle your mind. The food you eat, paper you write on, pencils you write with, your furniture, clothes, and the pleasant sights they bring are green gifts given from Hashem with great love.

CHAPTER 13

MINI MOUTHS

Since the Creator, may He be exalted, has free will in whatever He does, is neither compelled nor forced, and has no given nature, He created diverse things, according to the dictates and timing of His wisdom, so that their diversity would indicate His oneness and free will in all His actions, as it says: "Whatever God willed, He did—in heaven and on earth."

(Chovos HaLevavos, chapter 1)

See if you can guess who we are talking about? *It's a beautiful, bright day, the sun is shining, and you take a breath of fresh air. It's great to be alive, with another chance to grow in this world and reach for the sky.* Although we could very well be talking about a *yeshiva bachur*, we are actually talking about a tree!

You're familiar with how trees grow tall in order to get more sunlight, but did you know that they also breathe? Trees take in air using tiny holes on their leaves we call **STOMATA**. Air provides leaves with the main ingredient they need to make sugars, which is **CARBON DIOXIDE**. Like little microscopic mouths, the stomata open when the sun is shining and it's time to eat. At night, when the sun goes down and they can't make food, they close up to prevent unnecessary water loss.

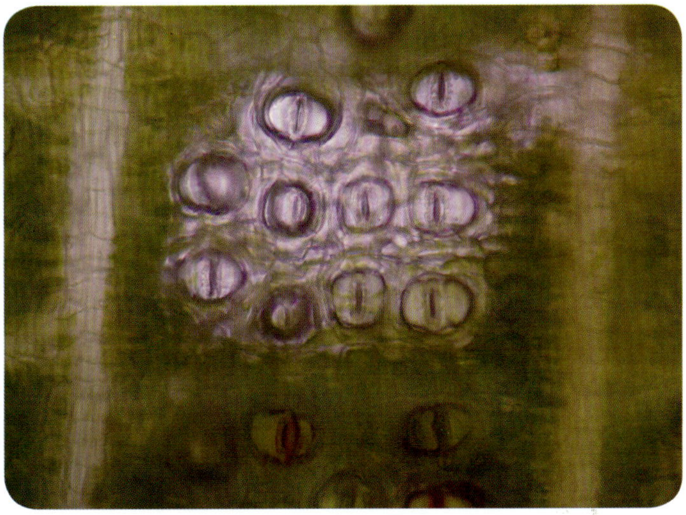

Did you ever imagine there were mini mouths on leaves? (Credit: Bigstock)

The marvel of how Hashem made trees able to eat air gets even more impressive when we realize how much people gain from this process. We, and all other animals, need **OXYGEN** to run a complicated process that transforms the energy contained in the food we eat into a form of fuel that our cells can use to build bones, grow hair, and lots of other things that our bodies do. When sugars are produced within leaves, oxygen is also made as a by-product and released into the air for us to catch when we breathe.

If you want to see phloem in action as it brings sap down a tree, just stick a tap in a maple tree and you will get some really yummy syrup.

Once sugars are made in leaves, they get sent to the rest of the tree in a solution of water and sugar we call **SAP**. Hashem designed special cells called **PHLOEM** to transport the sap, and they are located just under the surface of the tree. When phloem gets old it is pushed outward by new cells and becomes the outer bark of the tree. Even though it isn't alive, the old bark is still very important as it protects the living cells beneath it from damage by weather, insects, animals, and organisms that can cause disease.

If we go deeper into the tree, we will find another type of cell we call **XYLEM**, which is designed to bring water and nutrients from the ground up to the rest of the tree. They are the last living parts of the tree we will find, because underneath them is wood, which is made up expired xylem. Again we see the dead cells helping out as old xylem is used to make more wood in the heart of the tree, providing the support needed for it to grow taller.

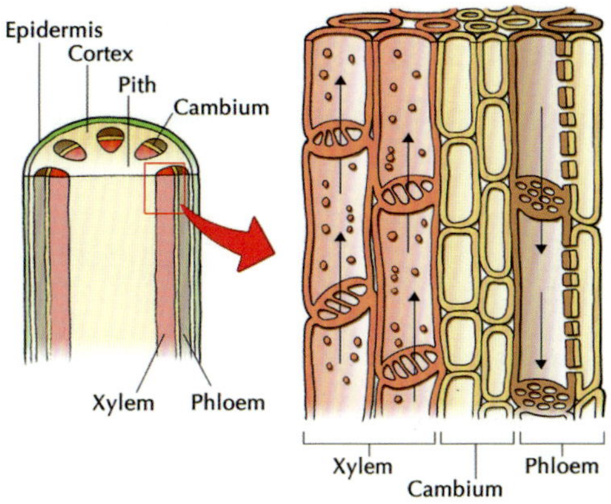

Have you ever wondered how old a tree is? When the tree is standing, it's pretty hard to tell, but after it's been cut down it's quite easy. A series of rings records a lot of information about the life of a tree. Wide, light lines record the rapid growth of the tree in the spring, whereas thin, dark lines record the slow growth of the winter. Since each ring represents a season, by counting them you can calculate the age of the tree.

See if you can tell how old this tree was before it was cut down.

Whether it is how they eat air, provide us with oxygen, make xylem and phloem, or have really neat rings, one thing is clear—trees are truly impressive and testify to the wisdom and kindness of our Creator.

CHOCHMAH V'CHESSED

Getting plants to open their mouths when they are hungry is no simple task. They don't have muscles in their mouths like we do to help them eat, or even brains that can remind them when they're hungry. Lining their stoma are uniquely designed cells called **GUARD CELLS** that react to sunlight by allowing water to fill them up. The increase in water changes the shape of the guard cells, effectively opening the stoma. When the kitchen is closed, like at night, they get the cue and release water, which makes them shrink back to a shape that keeps their mouths closed, and avoids unnecessary water loss.

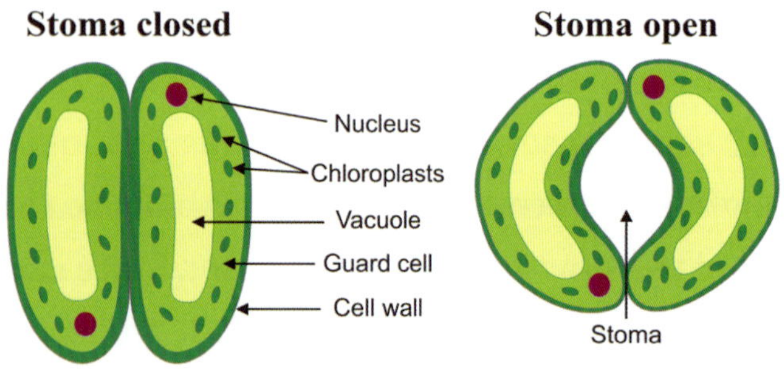

These guard cells remind me of electric doors that have sensors and open automatically. Using light sensors and hydraulic power, on top of the fact that they are microscopic in size, Hashem's automatic doors are the most amazing ones on earth. In fact, all the presents we get from plants rely on the mini-mouth design.

CHAPTER 14

SPECTACULAR SEEDS

In truth, Bnei Yisrael must know that miracles and nature are one and the same, and in truth, there is no miracle greater and more wondrous than that of nature, which is the greatest of all wonders that we can comprehend.

Sefas Emes (Parashas Behar Bechukosai תרל"ז)

Pods growing from a cocoa tree whose seeds will be used to make chocolate

Ever wondered which fruit is the biggest? It's a pumpkin and they can grow to be over two thousand pounds!

Can you imagine a world without fruits? What fun would eating be without them? No mouth-watering melons, tasty tomatoes, delicious dates or—the unthinkable—life without chocolate! While Hashem is super nice for giving us fruits and all the benefits they bring, what do the plants get in return for all the work that goes into producing them?

In order for a baby tree to grow properly, it needs to have a lot of water and sunlight. If plants were to just drop their seeds into the ground underneath them, their **SAPLINGS** (baby trees) would have to compete with them for water and light. Like all good parents, plants want to ensure their children succeed in life and so they have ingenious ways of distancing their seeds from themselves.

The most common means of seed dispersal Hashem made was to give them a very yummy covering we call fruits. Animals eat the fruits and throw out the seeds, which are inedible and are often protected by a very

This monkey thinks he's just having a snack, but he's about to plant a new tree.

hard casing. This enables the seeds to be taken far away from their parent plants, where they are likely to find a place with good soil, enough water, and abundant sunlight for them to grow.

Other plants make seeds like nuts and sunflower seeds that don't have fruit or have very hard shells. Although these seeds are food for many creatures, they often get planted unintentionally. Birds often drop them by accident while trying to fly them back to their nests. Animals like squirrels and mice collect and store seeds in holes under the ground and then sometimes forget where they hid them. Even ants are helpful planters, as they bring seeds underground to eat the nutrient-rich coating that surrounds them. Ants then throw out

"Hmmm, where did I leave all those seeds I buried?"

the seeds in one of the rooms they make for garbage—placing them in the perfect place to grow.

It gets even better: Hashem actually made flying seeds! Some plants (like dandelions) have a parachute-type structure attached to the seeds that enables them to get picked up by the wind and carried vast distances before landing. Perhaps even more spectacular is the seed of a maple tree, which has wings and flutters away from the tree like a helicopter when it falls. Last, but not least, is the seed bomb. Pea pods actually explode when the seeds are ready, and shoot the peas a good distance away from the trees.

Dandelion, maple seed, violet pod about to explode, and the chestnut with hooks that latch onto the hair of animals passing, who carry it along, so it can find a nice place to grow

Once under the ground, seeds still have some work to do before they sprout their first leaves. Until they get wet, seeds are in an inactive state called being **DORMANT**, which can last for years. When it rains, the seeds wake up and start to grow. While they can't make their own food yet because they lack leaves, Hashem gave them just enough food stored within their seeds to grow roots that can get water, and they then sprout a little leaf above the ground.

I can think of so many foods I enjoy that come from fruits, seeds, beans, and nuts and am so amazed by the ways they are distributed that you could call me a fruit fan, but to tell you the truth, I am an even bigger fan of the One who made them.

CHOCHMAH V'CHESSED

It is interesting to think about how seeds were likely the first form of life ever created. While the Torah teaches us that plants were created on the third day, *Rashi* explains they didn't actually come out of the ground until the sixth day, when they were watered for the first time. This would lead us to assume that the plants Hashem created on the third day were seeds, which are designed with a device that can sense when water is present. This is crucial for their survival because if they were to sprout when water wasn't present, they wouldn't last too long.

The ability of seeds to stay "sleeping" or dormant for years at a time helps plants to survive just about anywhere. While other forms of life could perish in harsh droughts due to a lack of water and food, plants will carry on, their "seeds" safely hidden under the ground waiting for better times before they sprout.

While the wisdom we find hidden in seeds is incredible, the kindness is impressive also. Walnut, pistachio, and sunflower seeds make tasty treats, and eating them can be so much fun. All these wondrous traits and more are found in Hashem's spectacular seeds!

CHAPTER 15

SECRET WARS

Rav Shach, zt"l, in one of his many discourses on emunah—after expanding on and detailing many of the wonders and Divine providence found in nature, and how we can see Hashem Yisbarach with our sensations completely—added, "I think about these things all the time."

Yalkut Ohel Rivkah by Rabbi Yosef Simcha Klein

Far out in forests, deep down in the jungles, there are, as we speak, secret wars being fought. They are so secret that even people passing by might not realize they are standing on a battleground. You are about to learn about the hidden struggle of plants to survive among a host of animals and insects that are out to eat them.

Plants have an obvious disadvantage from their predators, as they have no claws or teeth to fight back nor feet to run away. In place of the traditional tactics of self-defense, Hashem gave them marvelous means of fighting off their enemies. Some of these you are probably familiar with and others will blow you away!

As the woody parts of plants are well protected by bark or waxy coatings, it is usually the leaves that are of interest to **HERBIVORES** (animals that eat plants). How do they survive?

- Many plants have a strategy of growing massive amounts of leaves so that although many are eaten, they still have enough to produce the food they need.
- Others are less welcoming and have thorns or spikes that grow all over them and act like a barbed wire fence to keep out uninvited guests.

The razor-sharp spikes on this plant will make most hungry animals think twice about eating its leaves for lunch.

While razor-sharp projections can often deter hungry animals, they don't do much to keep away bugs. Insects are perhaps the greatest threat to most plants as they can come in swarms and have the potential to eat up everything green in their path. To combat flying invaders, plants wage chemical warfare against them.

This plant is in big trouble but it may have a surprise or two for these bugs, hidden under its leaves.

It's amazing: the pleasant smells of plants like peppermint, rosemary, or cinnamon are actually deadly toxins for insects. Fortunately for us, many of the chemicals plants produce to keep pests away are harmless to us and can even be tasty. In fact, most of the spices in your kitchen cabinets were once used as some form of bug spray to protect the plants they came from.

Another tactic used by plants in their battles with bugs is storing poisons in their leaves. An example of this is the bitter-tasting tannins within tea. These tannins disrupt the digestive system of insects who eat them. If the bugs eat enough tea they

Did you ever imagine the spices we put on foods are really bug sprays?

can die, but even if they survive a snack, they will get an awful stomachache and find other sources of food in the future.

Perhaps the most impressive ability plants were given to defend themselves is that they can tell their friends to prepare for an attack. Many plants only produce large amounts of toxic chemicals after their predators begin to eat them. While those on the front line may suffer a big blow from

What do you think these trees are talking about?

the bugs, other plants in the area may not get hurt as bad. Plants being eaten often send out messages to inform their comrades of the assault so they can pump up their poison production and be in a better position to fight off their foes.

If talking trees sounds funny, you might be surprised that their language was known by our rabbis long ago. The Gemara (*Sukkos* 28a) lists the praises of Rabban Yochanan ben Zakkai. Among his accomplishments was that he knew the language of trees. While I surely don't understand their leafy language, they seem to be saying one thing loud and clear: Hashem is awesome!

CHOCHMAH V'CHESSED

The fact that Hashem made a bug repellant that is healthy for us and makes our foods taste better is a sure sign of Divine design. Really, all the forms of self-defense that plants have are amazing. Just imagine trying to show someone who couldn't walk, or even move, how to fight off a host of hungry creatures out to eat him! Well, Hashem can do anything and we are lucky that we can appreciate and marvel at the astonishing ways He designed to provide every creature with its needs.

The benefits we get from the chemicals within plants go far beyond spicing up dinner. Oils from leaves, roots, and other parts of plants can be extracted and are called essential oils. They have a host of wonderful properties that include treating infections, freshening the air, making all-purpose cleaners, boosting the immune system, and releasing tension. The "secret" is out of the bag and we should all be extremely impressed at the amazing powers Hashem put into plants.

CHAPTER 16

FANTASTIC FUNGI

Mah rabu maasecha Hashem kulam b'chochmah asisa malah ha'aretz kinyanecha—How great are Your deeds, Hashem, all of them You have made with wisdom, the earth is full of Your creations.

(Tehillim 104:24)

What is the largest living thing on earth? It's not a blue whale or giant redwood tree, but a really big **FUNGUS**: In the soil of Oregon's Blue Mountain Forest lies a honey fungus that is over four square miles long (about 1,665 football fields) and weighs over a hundred tons! Don't worry, this monster won't eat you as fungi (plural for fungus) prefer to eat things that have already died.

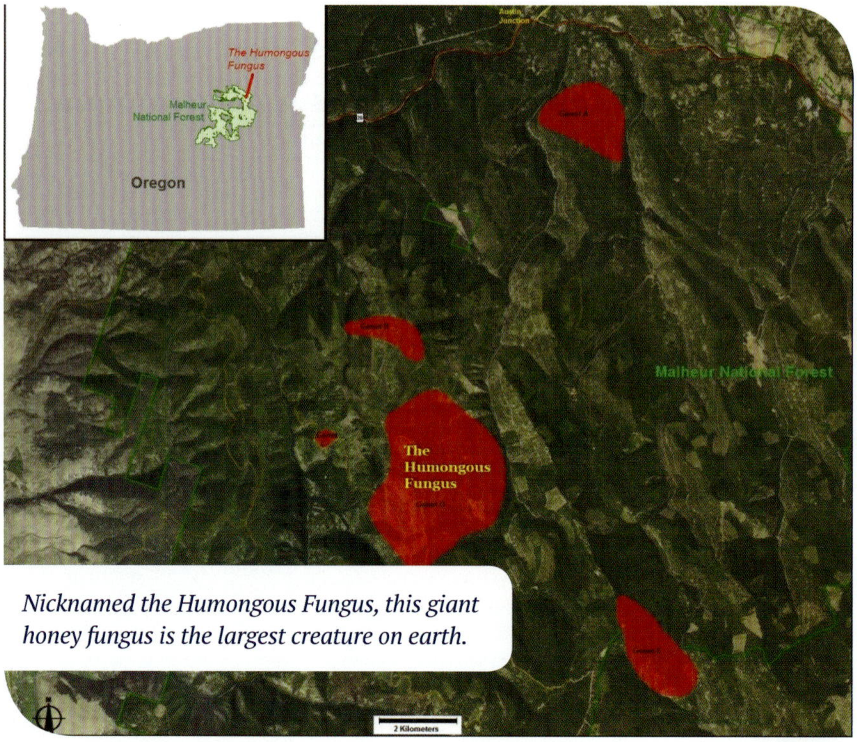

Nicknamed the Humongous Fungus, this giant honey fungus is the largest creature on earth.

How a fungus eats is pretty interesting, as it doesn't actually have a mouth. The body of a fungus, called the **MYCELIUM**, is usually found underground and is made up of a network of threadlike structures called **HYPHAE**. When fungi find food, their hyphae secrete digestive juices onto it. After it is broken down, the nutrients are absorbed directly into their bodies. The unique eating ability of fungus enables them to grow on and eat just about anything and helps them carry out what is perhaps their most important job—taking out the trash.

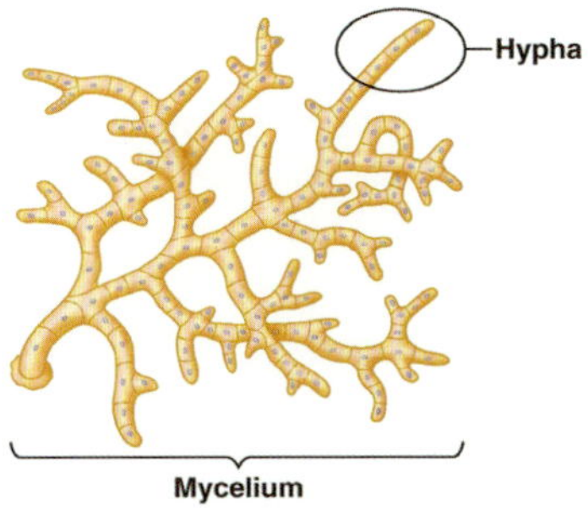

As fungi get bigger, they form mats of many hyphae called mycelium.

I like to think of fungi as the garbage men of the planet because whenever something dies, it is eaten and disposed of by fungus. In a relatively short amount of time, the dead animal or fallen tree vanishes, ensuring the world stays clean and beautiful. Another benefit of their eating is how fungi help **RECYCLE**. By breaking down what they eat into microscopic pieces, they enable those nutrients to be used again and help new life grow.

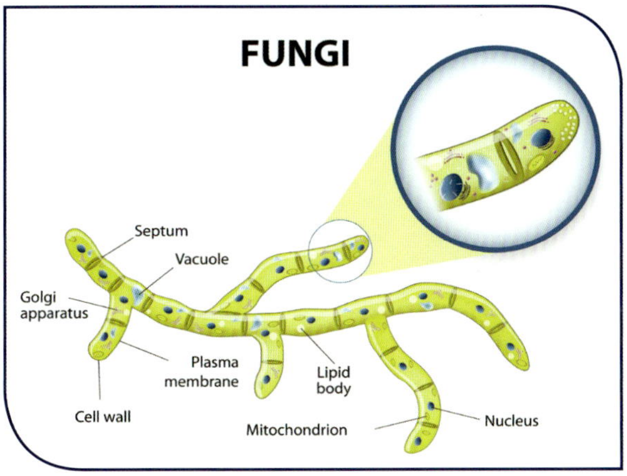

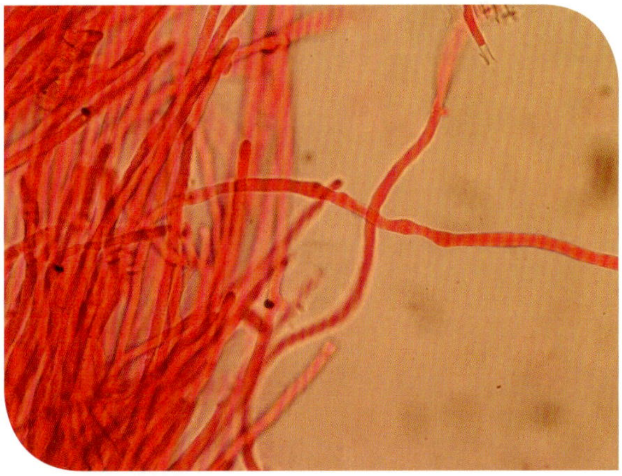

Fungi are made up of little "arms" called hyphae.

Fungi are able to find things shortly after they die with the help of **MUSHROOMS**. Mushrooms are sort of like the "fruit" of fungus in that they have thousands of microscopic "seeds" we call **SPORES** that are capable of

Let the spores soar!

growing into a new fungus. They are released into the air and fly around until they find something to grow on like a dead tree or piece of meat you forgot to take out of your lunch box.

Giant puffball mushrooms like this one are known to explode, sending a white cloud of trillions of spores into the air.

Known as the world's largest mushroom, the macrocybe titans is safe to eat and makes a great toy for kids living in Africa where it grows.

Up for a tasty treat? Beautiful boletus campestris mushrooms like this one are known to be delicious. But beware, they have cousins that look like them—but are poisonous.

Unfortunately, when most people think of fungus they associate it with itchy feet, rotting food, and moldy walls. The reality is that the benefits it brings us far outweigh the trouble it causes. For instance, about eighty percent of all plant life needs a type of fungus that lives on their roots called **MYCORRHIZAE** in order to grow properly. The bodies of mycorrhizae, which are made up of huge mats of hyphae, can extend two hundred times further than the roots of trees, enabling them to get much more water and nutrients from the soil than they would

Did you ever imagine there were mushrooms that glow in the dark? The mycena chlorophos mushrooms are Hashem's flashlights in the forest.

If your pen is low on ink, you may want to look for an ink cap mushroom for a refill.

without their help. In return for the favor, plants send the fungus a portion of the food they make.

So, the next time you see slimy stuff growing on your walls or weird-looking growths on food you forgot to throw away, after being grossed out, you may want to take a minute or two to think about how Hashem made such a fantastic thing we call fungus that helps us in so many ways (like keeping forests clean, providing us with tasty treats, making medicines, and more).

CHOCHMAH V'CHESSED

It is really mind-boggling when you think of how great a gift fungi are. The green fuzzy stuff you may have found annoying and gross is actually a growing collection of hyphae that is breaking down, cleaning up, and recycling the garbage you forgot to throw away. Pretty amazing! Hashem made garbage men that work 24/7, are located on every square foot of our planet, and can recycle all the dead or decaying matter on earth, in addition to doing a great cleanup job!

To appreciate how impressive a creation fungi are, try to imagine a great contest proposed to scientists and engineers to design a worldwide cleaning and recycling service. It would require a tremendous amount of time, effort, and money, but world leaders felt it was worth the investment. Now just think of what they would say about someone who designed cleanup crews that didn't cost anything to produce, were able to travel and work anywhere on earth, and could quickly and efficiently recycle the garbage on location.

In addition to help cleaning up fruit that was left out, this penicillum mold has saved hundreds of millions of lives by way of the antibiotic penicillin that comes from it.

If you think such an accomplishment is worthy of praise, make sure you say "*Mah rabu maasecha Hashem kulam b'chochmah asisa*" with extra *kavanah* in your next *tefillah*! Hashem's design of fungus is so much better than merely cleaning and recycling the garbage of our planet, as they also make a host of helpful things like medicines, bread, and wine, to name a few. Add to the list how most plants need fungus to survive and thrive, and we definitely have to thank Hashem for the fantastic fungi He gave us.

Did you ever wonder why we make a *berachah* of *She'hakol* and not *Ha'adamah* on mushrooms? Seeing as they grow out of the ground, the logical blessing would be the same as plants, *Ha'adamah*. This is such a good question that there is a discussion in the Gemara (*Berachos* 40b) about why the *berachah* on edible fungi (*kemeihin* and *pitrios*) is *She'hakol*. The Sages of the Talmud had a profound knowledge of the nature of fungi and understood they were very different forms of life than plants. This was clear to them because fungi's primary source of nutrition clearly doesn't come from the ground, like plants, as can be seen from the fact that many mushrooms grow on top of trees and stones (*Meiri*).

Some kemeihin (truffles) can sell for over six thousand dollars and are worth way more than their weight in gold.

There is a bit of debate as to what the difference is between *kemeihin* and *pitrios*, although everyone agrees they are two different forms of fungi. Some Acharonim explain that *pitrios* are the familiar mushrooms we are used to eating, and grow above the ground, whereas *kemeihin* are mushrooms that grow under the ground and are a delicacy we call truffles today.

The fact that the Sages of the Talmud understood the nature of these fantastic fungi as being different from other plant life that grows out of the ground is remarkable because until recent history, scientists always considered fungi a form of plant life. I can imagine it being difficult for some Jews in the past, fluent in both the Talmud and science, to make the blessing of *She'hakol* on mushrooms, seeing as the scientific community determined they were a form of plant life, contrary to the conclusion of the Talmud. This conflict, like all those we find between Torah and science, just required the scientists time to gain a better understanding of the natural world until they came to realize the Sages were right and fungi are unique creatures, very different from the plants they share the ground with.

CHAPTER 17

FABULOUS FLOWERS

Avraham Ibn Ezra wrote in his poem, "God is Everywhere": Wherever I turn my eyes, around on Earth or to the heavens / I see you in the field of stars / I see You in the yield of the land / in every breath and sound, a blade of grass, a simple flower, an echo of Your holy Name.

What is your favorite kind of flower? A ruddy red rose, dazzling daisy, or maybe a stunning sunflower? While we may disagree about which type is the best, the fact that flowers are one of most fabulous forms of vegetation on the planet is undisputed. Can you think of any part of a plant that brings so much joy to so many people as flowers do? Whether for cheering someone up, decorating your Shabbos table, or given as a gift for a special occasion, flowers always seem to be out there making people happy.

Tulips hold the record for the most expensive flowers ever sold. During tulipmania in the 1620s, they were worth more than their weight in gold.

With all the great uses and delights we get from flowers, it is easy to see why Hashem made them for us, but what do plants get out of making such fancy features? While a plant may have many wondrous traits, being a *baal chessed* is not one of them; it wouldn't waste the energy and resources needed to produce flowers if it wasn't getting something important in return.

In order for plants to make seeds, they need a yellow powdery substance we call **POLLEN** that generally must be obtained from other plants. Seeing that they can't walk over to their friends to ask them for some pollen, plants were designed with amazing ways of sending and receiving the pollen they need. Some make huge amounts of pollen and release it into the air with the hope that a bit will land on the appropriate plant. In the spring, there is so much pollen in the air that a lot of it ends up in your nose and is notorious for causing an allergic reaction called hay fever among some people. While wind pollination works, the most common and effective form of transferring pollen actually involves the use of flowers.

If pollinating were a competition, bees would get the gold medal, as they carry the pollen for over half of all plants on earth.

The bright and beautiful colors of flowers are not only attractive to you—many creatures we call **POLLINATORS**, including various types of insects, birds, and bats find them fascinating. Flowers want to be found and many even give off a sweet smell to let pollinators in the area know they are around. Once in sight, pollinators can't resist flying over to see what the flower has inside. When they land, they find a sweet treat called **NECTAR** waiting for them in the center of the flower. As the pollinators are feasting away, their feet or mouths get a bit dirty because the center of the flower is lined with pollen that sticks to them. When they fly to other flowers for more food, the pollen rubs off and is effectively transferred between plants.

Did you ever imagine a flower could be so big?

The effort plants put into making flowers and nectar pays off as they acquire for themselves a host of pollinators that work for them in exchange for some food. The happy ending of this story is how so many creatures benefit from Hashem's brilliant invention of flowers. Plants get the pollen they need to make seeds, the pollinators get a sweet treat, and we get to enjoy their sweet aroma and fabulous display.

CHOCHMAH V'CHESSED

If you ever want to run a successful business, everyone will agree that you need good marketing. After all, if no one knows about what you're selling, you won't have too many customers. Flowers are the marketing tools Hashem gave plants to attract customers, usually insects that help pollinate plants. Hashem knows best what works in marketing, as beautiful colors and irresistible aromas leave flowers buzzing with bees.

Flowers have a special ability to make people happy.

Another element of Divine wisdom found in flowers was revealed by scientists conducting research at Rutgers University. They discovered that flowers improve people's emotional health and concluded that the presence of flowers triggers happy emotions, heightens feelings of life satisfaction, and affects social behavior in a positive manner far beyond what was normally believed. Only Hashem could put such amazing power in a flower!

CHAPTER 18

DESIGNER HOMES

The scriptures (Tehillim 114:18) assert: "God created the high mountains for the wild goats," not, "God created the wild goats on the high mountains," teaching us that God created the mountains for the sake of the wild goats. The pasuk continues, "the rocks as refuge for the gofers," not, "gofers take refuge in the rocks." God created the rocks for the sake of the gofers.

(Bereishis Rabbah 12:9)

Have you ever dreamed of traveling around the world? Just think about all the interesting people you might meet or the amazing things you would see. While nobody knows what types of adventures you would have on your trip, the types of natural scenes you would see is quite predictable.

As any world traveler will tell you, the types of trees, animals, and climate stays more or less the same for a long time and then, all of a sudden, starts to change drastically with strange new creatures everywhere. Scientists refer to the large areas of the planet that share similar features as **BIOMES**. On land, there are four basic ones:

- **DESERT** biomes are characterized by having very little rain. The temperatures are extremely hot during the day and then get very cold at

Camels are incredibly suited to life in the desert as they store food in their humps, have clear eyelids for protections against sand, and can go up to six months without drinking water.

night. Only a hardy bunch can survive life in the desert, such as mountain lions, golden eagles, and scorpions. Most plants can't live with such little water, but cactuses do pretty well on account of their specially designed leaves that prevent water loss.

For monkeys who love to climb, swing, and hang on trees, forests are the most fun.

- On the opposite extreme is the **FOREST** biome, which is often teeming with life. Tall trees make up the "roof" of forests, with a dense covering of leaves we call the **CANOPY**. In the shade below lies a second layer of shorter trees and vines called the **UNDERSTORY**. All kinds of animals live in forests like monkeys, tigers, and a host of birds who find all the food and water they need to survive and thrive.
- Somewhere in the middle, between deserts and forests, lies the **GRASSLAND**

For vegetarians like these, grasslands are great, as there is always something around to eat.

biome. These areas get rain for about half the year but then dry out for the summer. Poor soil and frequent fires keep the number of trees around low but make the perfect place for grasses like wheat and corn to grow. The classic grazing animals like wildebeests, elephants, and horses are found here along with the predators like lions and wolves that eat them.

- The coldest of all the biomes is the **TUNDRA**. Not much grows here as the soil, called **PERMAFROST**, stays frozen all year long. For only a short time in the summer, temperatures get warm enough for the top few inches of ground to thaw a bit and allow some small plants to grow. A few animals live in the tundra year-round, such as the polar bear and arctic fox, but in the summer, many migrating animals like caribou and snow birds come for a visit.

It is quite amazing how animals are so well suited to life in their respective biomes. Many have wondered how this came to be. David HaMelech actually solved this riddle a long time ago, as it says in *Tehillim* 104:18: "(He) made the high mountains for the wild goats and the rocks as refuge for the gofers." Hashem designed the world so that each animal He wanted in His world would find a home outside, just right for their particular needs.

Polar bears don't mind living in a freezer, and with their thick fur coats, even find it quite cozy.

CHOCHMAH V'CHESSED

Certain people love the outdoors while others are city folk. Just like we have our particular preferences of what type of place we like to live, animals do too. The polar bear has his winter wonderland and the vulture has its dream desert.

In addition to being suitable homes, the various biomes were created in order to provide for our needs, each in its own unique way. For example, treasures are buried beneath deserts, such as vast reservoirs of oil that we tap into for fueling our cars, homes, and electronic devices. Forests have a vast variety of plants, many of which have healing properties. Grasslands are where most grains grow and are the bread baskets of the world. The icy tundra is a summer getaway for all kinds of birds who use it as a safe place to raise their young, making it a key player in the entire world's food chain.

We listed only four basic biomes, but many scientists break up the earth into nine. When we think about biomes, we can appreciate the Divine planning and design that went into housing the insects, animals, birds, and reptiles that live on land—and we haven't even begun to talk about the sea!

CHAPTER 19

THE FANTASTIC FOUR

To everything there is a time and a season under heaven.
(Koheles 3:1)

Do you find it annoying to get all bundled up in the winter? Do you dread the blistering heat of the summer? With all the discomfort that goes into keeping up with the seasons, what do you think Hashem had in mind when He made them? Let's take a closer look at the seasons and hopefully we'll find some answers.

With all the fun you can have with snow, winter is sort of like a huge toy store where everything is free.

- During the **WINTER**, we get a lot of rain and snow. While making snowmen might be fun, the main benefit of this season comes with the abundance of water that fills up the lakes, rivers, and underground aquifers that were designed to store water for us throughout the year. Believe it or not, winter is also the greatest form of pest control on the planet. Insects can't do much in the cold, so their numbers go way down in the winter and ensure their populations don't grow out of control.
- Following the winter is the most colorful season of all, **SPRING**, that takes advantage of the abundance of water to get things growing. It is pretty easy to appreciate the benefits of spring

Do the colors and beauty of spring help you feel refreshed and alive?

Do you like having fun in the sun?

when green leaves return to the trees, flowers are in bloom, and everywhere is buzzing with life.

- Once the rain is long gone and the sun is out in full force, **SUMMER** has arrived. Cooling off in a pool (or with a good water fight) are almost essential

For you, fall may be your favorite, as it is so serene.

to enduring the heat. Although having fun in the sun may be the highlight of this season, the high temperatures and lack of rain actually enable us to get the majority of the food we eat: Grains like wheat need to dry out before they are ready to make things like bread, cereal, and noodles. A nice hot summer is just what they need to get ready…for your lunch.

- Last but not least is the **FALL** in which shorter days and less sunlight let the trees know it's time to get ready for the bitter cold of the winter. They shed their leaves to conserve energy as their food supply—sunlight—is getting slim. Many birds and other migrating animals also get the hint and travel in fall to warmer places to wait out the winter.

There are many more fantastic benefits of the different seasons, but before we end, I have to tell you one more which I always found amazing. It is how the seasons help sustain aquatic life in ponds and lakes. The system is so incredible that only Hashem could have thought of it. In order for life to endure, nutrients have to be recycled. When things die in lakes, they sink to the bottom. If it stayed warm all year long, those nutrients would be lost forever as dense cold water at the bottom would never mix with the lighter warm water up above. Seasonal changes get the water churning and bring the nutrients from the bottom to the surface where fish and other aquatic creatures can use them!

CHOCHMAH V'CHESSED

Some benefits of changing seasons are obvious, like how boring things would be without winter snowball fights or water sports in the summer. Still, while a change is often healthy, the wisdom Hashem had in mind when He made the different seasons is far more profound.

If not for the summer coming on time, the sea ice we learned about in Chapter 8 would keep on growing and many creatures living near the chilly seas couldn't survive. And if it was summer all year long, there would be no sea ice and all life in the sea would be lost without having any motion in the bottom of the ocean.

It isn't only us and the animals who benefit from changes in seasons—many plants are picky about when they like to grow their fruit. Fruits like cherries, figs, and mangos grow best in the summer while apples, bananas, and pears ripen in the winter. All life was designed to have its season, its time to shine. We can see the great plan and purpose, wisdom, and kindness that fills every place and every "time" in our world.

CHAPTER 20

SUPER SHIELDS

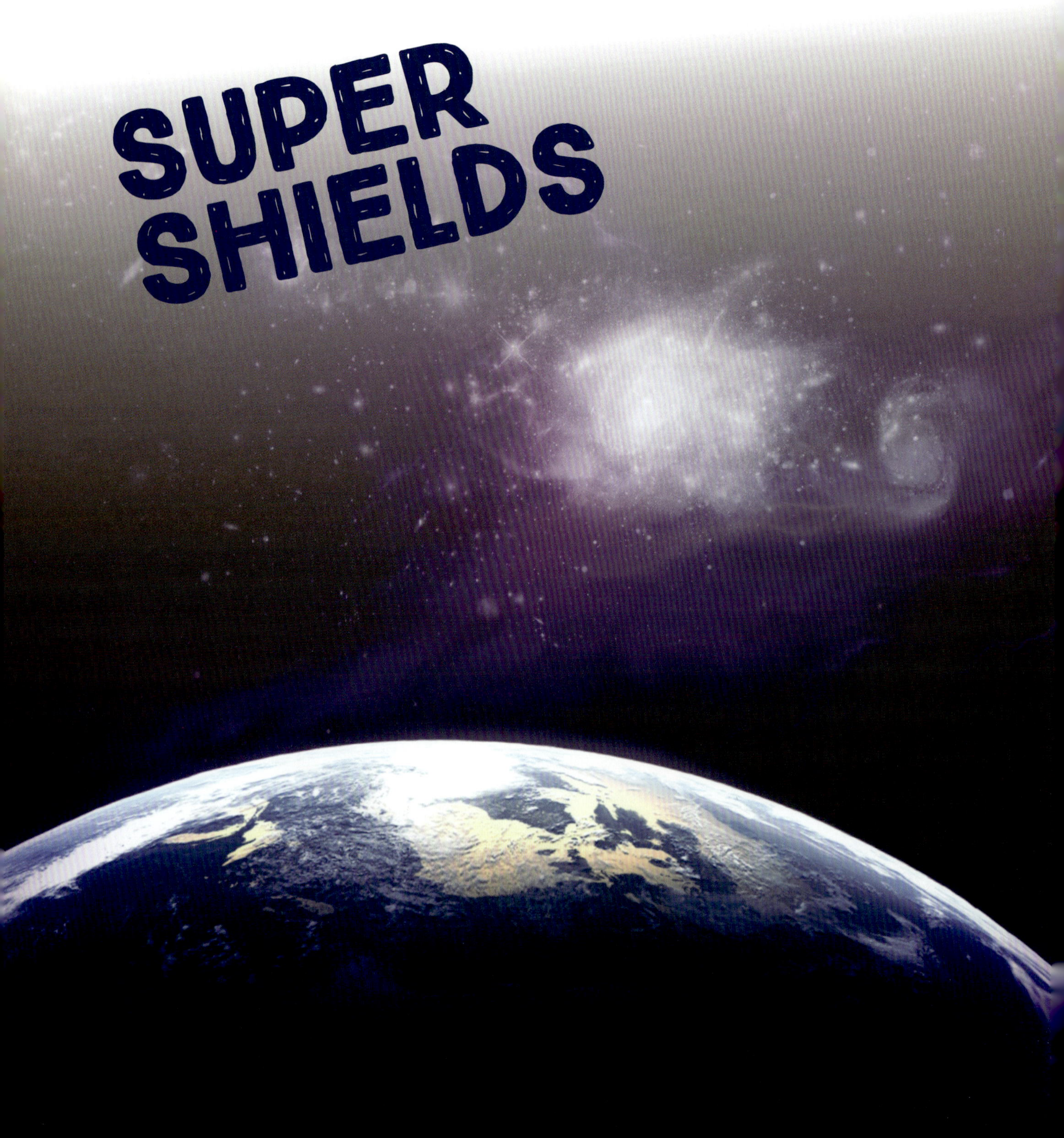

The heavens recite the glory of God, and the sky tells of the work of His hands.
(Tehillim 19:2)

Did you know that at this very moment you are under attack! A stream of deadly cosmic rays is flying toward our planet from the sun. Don't panic, there is no need to get scared or head for a bunker—Hashem gave us a pair of super shields to protect us.

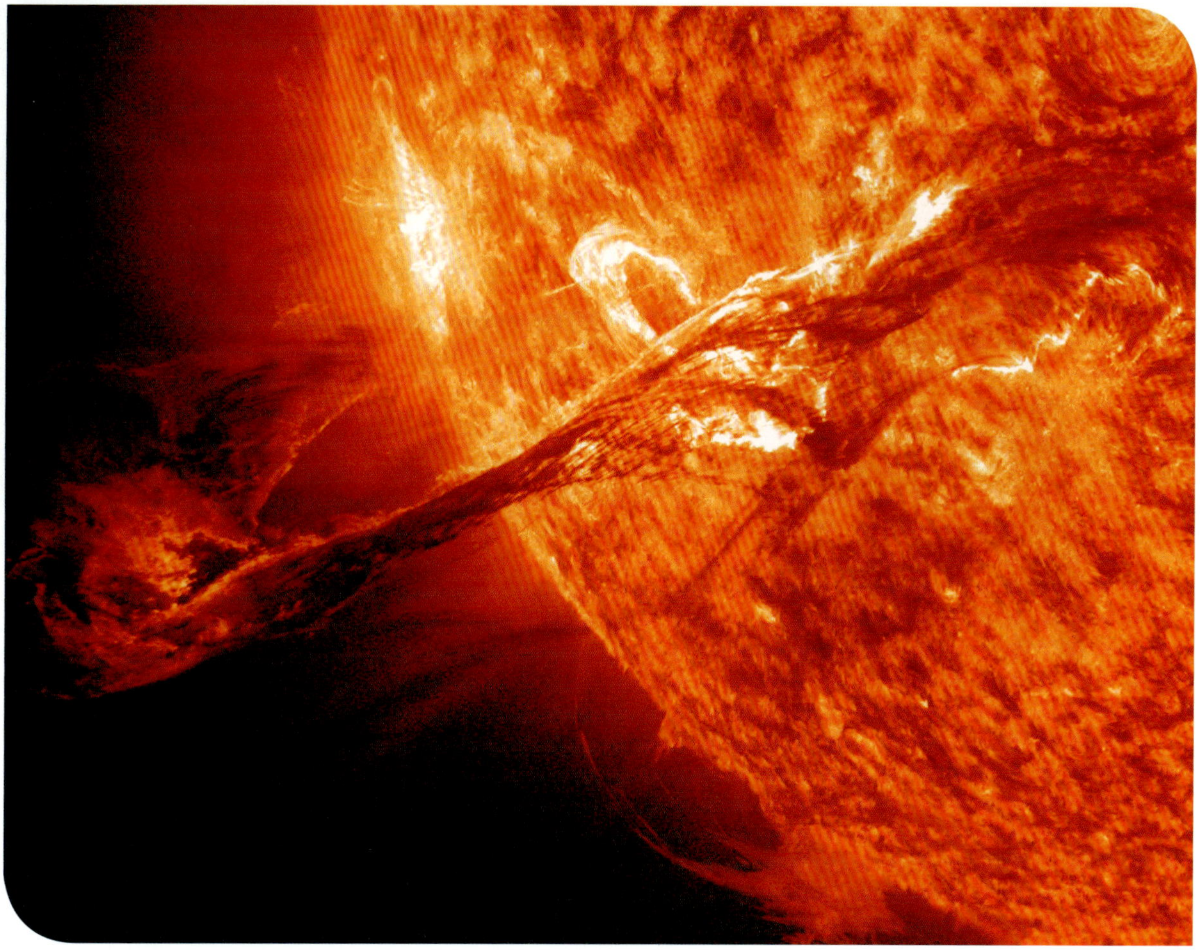

Giant explosions on the surface of the sun would be devastating if not for the super shield Hashem made to protect us.

The first one you may have heard about is the **OZONE LAYER**. Ozone is a form of oxygen gas less common than the type we normally breathe. About 20 percent of the air around you consists of oxygen atoms that like to travel together in pairs. Way high up in the sky, these pairs get broken up by a harsh form of light coming from the sun we call **ULTRAVIOLET RADIATION** or UV rays. The lonely oxygen atoms often tag along with a pair that didn't get hit by a heat ray and make a trio we call ozone.

Ozone is different than ordinary oxygen in several ways. First of all, ozone is blue and smelly while the oxygen outside is clear and odorless, which is why you don't notice it hanging around. Another difference is that ozone has a special ability to prevent us from getting "fried" from the sun.

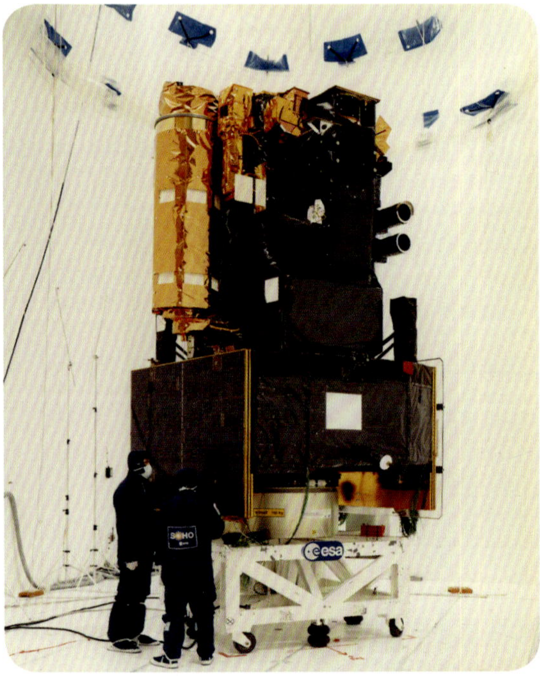

This spacecraft, called the SOHO, monitors space weather, including fluctuations in the solar wind and magnetosphere.

If you have ever stayed out in the sun too long, you probably got a sunburn and felt firsthand the danger of UV rays. The sun emits a lot of different types of light. The type we use to see things, called the **VISIBLE SPECTRUM**, is harmless, but others like ultraviolet light can cause a lot of damage. Seeing that the layer of ozone in the sky blocks about 98 percent of the UV light entering the atmosphere, you can only imagine how badly we would get burned if it wasn't there.

While the ozone layer does a great job of shielding us from ultraviolet light, there is a much greater danger lurking out in space. There is a constant flow of deadly particles escaping from the sun, called **SOLAR WIND**. If solar wind reached our world, it would likely result in massive power outages, cause widespread disease, and might even strip the earth of its atmosphere, making it impossible to breathe!

Don't worry, in order to protect us from this peril, Hashem gave us a really big magnet.

Yes, I said magnet, and the amazing powers of magnetism go way beyond the fun of playing with Magna-Tiles. Deep within the earth lies a liquid iron core. Steady currents

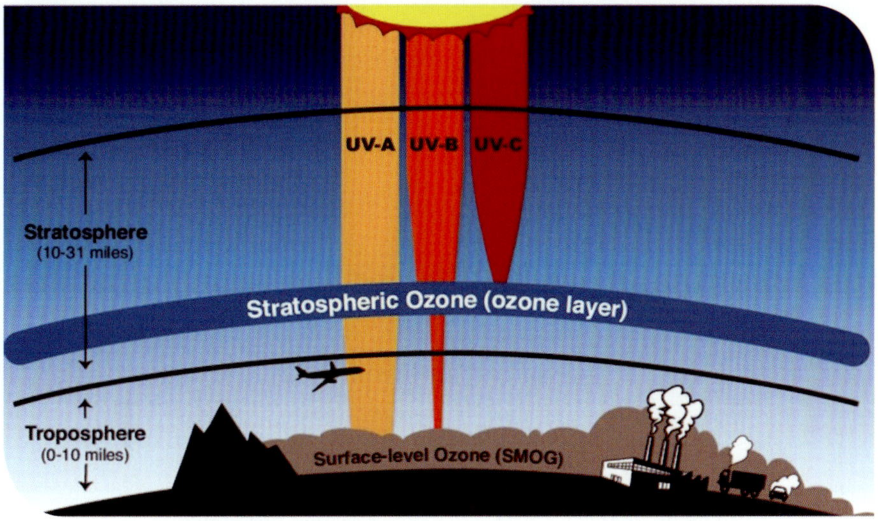

The ozone layer is literally the greatest sunscreen on earth.

of molten iron generate electric currents whose influence can be felt far beyond the earth in what we call the earth's magnetic field or **MAGNETOSPHERE**. Seeing that the solar wind is made up of electrically charged particles, when it gets near the earth it is deflected by the powerful electric currents of the magnetosphere that oppose it.

Some charged particles from the solar wind make it to our planet through "holes" in the magnetosphere at the North and South Poles and make what is arguably the most spectacular light show on earth.

Credit: NASA/GSFC/SOHO/ESA

Did you ever imagine there was such a big shield around our planet to protect us?

I don't know about you, but when I hear about invisible shields like the ozone layer and magnetosphere, I feel so fortunate to live in a marvelous world of wonders!

CHOCHMAH V'CHESSED

While a compass is pretty outdated with the modern GPS systems out there to help us find our desired destinations, it is never too late to praise Hashem for putting a giant magnet beneath us that has helped countless travelers find their way around the world.

For me, the magnetosphere is one of the most amazing wonders of Hashem in our book. I particularly like to think about how Hashem is and always has been protecting us at every moment of our lives. Even though we don't see the magnetosphere, it is constantly blocking all kinds of dangers from reaching us.

What is the lesson of the magnetosphere and ozone layer? Well, if you ask me, it's that Hashem is always watching over us and making sure no cosmic rays, thieves, or any other sort of harm reach us that aren't supposed to. The super shields Hashem made to protect us are really impressive, but our true Super Shield is Hashem and His protection is perfect!

CONCLUSION

It is truly amazing what happens when you begin to reflect upon the wonders of our Creator. A simple glass of water that was once part of the unnoticed background scenery of your lives can become a springboard to a multitude of reasons to thank and praise the One who designed and created the marvelous liquid you see. The uses of water are close to endless when you think about it. A healthy and refreshing liquid that cleans everything, from your body to the floors. It helps break down the vitamins, minerals, and nutrients you need and enables them to get to every cell in your body. Due to its "sticky" nature, trees can grow tall, and it turns into ice when it gets really cold to ensure the fish don't freeze to death in the winter. If that isn't enough to impress, what about the system Hashem designed to deliver twenty-three quadrillion gallons of water every year from the oceans to the land that runs continuously, requires no maintenance, and costs no money.

The wood on your table or paper in the book you are reading can really keep you connected to your Creator because wood is an awesome present given to our planet. How often and in what ways do you benefit from wood? In just a few minutes you will surely have a long list and when you start thinking about the wisdom found in trees, leaves, and flowers, you'll be overwhelmed in amazement.

How many times do you see metals in a day? Instead of looking at cold, dark iron, you can see the light of Hashem's wisdom revealed before your eyes when you think about how He put all kinds of useful materials in the earth below us for us to find and use. Big buildings, car engines, coins, jewelry, and a vast amount of other things we use every day just wouldn't be around if not for the gift of metals.

These are just some examples of how joy and delight await those who spend time contemplating their Creator. In the art gallery of our world, brilliant and beautiful masterpieces can be encountered with every step we take. By looking around at the marvels of the world and reflecting upon your daily lives, a multitude of people, places, and things can be found to enjoy and be grateful for. Even a poor man can feel truly wealthy with an awareness of the treasures of life he had overlooked in the past. May it be the will of Hashem that the end of this book becomes the beginning of your exploring the wonders of our Creator and that you receive much pleasure from the wisdom and kindness that you find.

GLOSSARY OF TERMS

ADIABATIC COOLING: The process of reducing heat through the lowering of air pressure, like what happens when air rises rapidly.

AIR PRESSURE: The force exerted by air on any surface in contact with it.

AQUIFER: A body of permeable rock that can contain or transmit groundwater.

ATOM: The smallest particle (unit of matter) of an element which can exist and be part of a chemical reaction (a source of nuclear energy).

CLOUD SEEDS: The common name of cloud condensation nuclei, or CCNs, which are the smallest particles in which water condenses in the formation of clouds.

DORMANCY: A period in an organism's life cycle when growth, development, and (in animals) physical activity are temporarily stopped.

EARTHQUAKE: A sudden and violent shaking of the ground, sometimes causing great destruction, as a result of movements within the earth's crust or volcanic action.

ELECTRON: One of the fundamental constituents of matter having a negative charge and found orbiting outside of the nucleus of an atom.

ESTUARY: A habitat created when the fresh water of a river meets the salt water of the ocean.

EYE WALL: The area immediately outside the eye of a hurricane or cyclone, associated with tall clouds, heavy rainfall, and high winds.

FALL: Also referred to as autumn, it is the third season of the year after summer and before winter, when crops and fruits are gathered and leaves fall.

FAULT: A break in the continuity of a body of rock caused from significant displacement as a result of rock-mass movement.

FUNGUS: (pl.–fungi) Any member of the group of eukaryotic organisms that includes microorganisms such as yeasts and molds, as well as the more familiar mushrooms.

FUNNEL CLOUD: A rotating funnel-shaped cloud forming the core of a tornado or waterspout.

GLACIER: A slowly moving mass or river of ice formed by the accumulation and compaction of snow on mountains or near the poles.

GROUNDWATER: Water held underground in the soil or in pores and crevices in rock.

GUARD CELL: Each of a pair of curved cells that surround the parts of leaves that take in air called stoma, which becomes larger or smaller according to the pressure within the cells.

HERBIVORE: An organism that eats only plants.

HYDROELECTRIC POWER PLANT: A power plant using water to generate electricity for distribution.

HYDROGEN BOND: A weak chemical bond responsible for the bonding of water molecules in which an electropositive atom (usually hydrogen) of one molecule is attracted to an electronegative atom of another molecule.

HYPHAE: Cells that attach to one another in long, string-like tubes that make up the body of a fungus.

ICEBERG: A large floating mass of ice, detached from a glacier and carried out to sea.

JET STREAM: A narrow belt of high-altitude winds (about 12,000 meters high) moving east at high speeds.

MAGMA: Molten material beneath or within the earth's crust, from which lava is formed.

MAGNETOSPHERE: A region surrounding the earth, beginning about sixty miles above the surface and extending hundreds of thousands of miles into space. In this region, the earth's magnetic field exerts a significant influence on any charged particles that encounter it, thereby deflecting most of the charged particles coming from this sun in the form of solar wind and protecting the planet.

MUSHROOM: The fleshy part of a fungus that bears spores, grows above ground, and consists usually of a stem bearing a flattened cap.

MYCELIUM: The body of a fungi, which is made up of a mass of many hyphae.

MYCORRHIZAE: A symbiotic association of plant roots and fungi in which both benefit.

NECTAR: A sugary fluid secreted by plants, especially within flowers to encourage pollination by insects and other animals.

NEUTRON: A subatomic particle of about the same mass as a proton but without an electric charge, present in all atomic nuclei except those of ordinary hydrogen.

OZONE LAYER: A layer in the earth's stratosphere at an altitude of about 6.2 miles (10 km) containing a high concentration of ozone gas, which absorbs most of the ultraviolet radiation reaching the earth from the sun.

PERMAFROST: Soil that is frozen all year long.

PHLOEM: The tube-like vascular tissue where food moves down from the leaves toward the roots in plants.

PHYTOPLANKTON: A diverse group of single-celled organisms that produce food for life in the water by way of photosynthesis.

PLATE TECTONICS: In geology, a theory that the earth's lithosphere (the crust and upper mantle) is divided into a number of large, plate-like sections that move as distinct masses.

POLLEN: The fertilizing element of flowering plants, consisting of fine, powdery, yellowish grains or spores.

POLLINATOR: An organism that brings pollen to a flower so as to allow fertilization.

PROTON: An atomic particle found in the nucleus, similar in size to a neutron but carrying a positive charge.

RECYCLE: To treat or process (used or waste material) so as to make suitable for reuse.

SAPLING: A young tree.

SEA ICE: Ice formed by the freezing of seawater.

SOLAR WIND: The continuous flow of charged particles from the sun that permeates the solar system.

SPORE: Durable, single-celled structures, capable of being transported by way of the wind or water, that give rise to new organisms after finding suitable conditions for growth.

SPRING: The season after winter and before summer, in which vegetation begins to appear.

SUMMER: The warmest season of the year occurring between spring and fall.

TORNADO: A localized, violently destructive windstorm characterized by a long, funnel-shaped cloud extending toward the ground and made visible by condensation and debris.

TRANSPIRATION PULL: The proposed mechanism by which trees draw water through their roots.

ULTRAVIOLET or **UV RADIATION:** The portion of electromagnetic radiation (sunlight) having a wavelength shorter than that of the violet end of the visible spectrum but longer than that of X-rays.

VISIBLE SPECTRUM: Light within the range of wavelengths to which the eye is sensitive.

VOLCANO: A mountain or hill, typically conical, having a crater or vent through which lava, rock fragments, hot vapor, and gas are being or have been erupted from the earth's crust.

WATER TABLE: The depth at which underground water is first encountered.

WATERSPOUT: A tornado occurring over water that forms a column of water and mist extending between the surface and the clouds above.

WINTER: The coldest season of the year, which occurs between fall and spring.

XYLEM: The tube-like vascular tissue where water and nutrients move from the roots up toward the leaves in plants.

ABOUT THE AUTHOR

Rabbi Yaakov Lubin is the co-founder and director of The Jewish Center for Science, which is dedicated to creating educational resources that show how the wonders of science and nature are truly the wonders of Hashem. Rabbi Lubin is the author of *Fundamentals of Life Science*, a Torah-based middle-school textbook for Jewish schools. It is the first volume in a series of textbooks called *Niflaos HaBoreh Explorer*, which present the sciences as the handiwork of Hashem while maintaining the highest level of secular education. He also gives a spectacular multimedia presentation on *niflaos haBoreh* called "Barchi Nafshi," which has been enjoyed by numerous middle schools, high schools, shuls, and nursing homes. For more information on bringing *Fundamentals of Life Science* to your school or the "Barchi Nafshi" presentation to your school or shul, Rabbi Lubin can be contacted at YaakovLubin@jewishcenterforscience.com.

The JEWISH Center For SCIENCE
DISCOVER THE DIVINE DESIGN

Founded by Andrew Sossin and Rabbi Yaakov Lubin

Thank You!

MR. YEHUDA SEINFELD

for sponsoring a chapter of this book
and for his support of
The Jewish Center for Science

Thank You!

MR. LEON BENLOLO

for his help and support
of The Jewish Center for Science

Thank You!

JOSEPH BENLOLO

for his help and support
of The Jewish Center for Science

Thank You!

MRS. LEAH TARLER

for her help and support
of The Jewish Center for Science

Thank You!

MRS. ROMELLE W. TARLER

for sponsoring a chapter of this book
in the memory of

Nathan A. Tarler

Nachum ben Liba

a proud Jew and fellow educator

Friends of The Jewish Center for Science

MAYER CHEMTOB

EZRA DEBBAH

JORDAN DITCHET

SHAUL DOUECK

ROBERT FINN

DAVID GILL

MARC HAMMERMAN

DAVID JEMAL

MICHAEL KAUFMAN

JACOB KASSIN

AHARON KURLANSKY

AVIYAH LEV

DR. ISAAC MADEB

YAAKOV MISHAN

RABBI YOSEF OZEIREY, SHLITA

ALEX PARADIZ

ALEXANDER POLTORAK

JORDAN RAYBOY

SHLOMO AND TAMAR RECHNITZ

RABBI SHMUEL ROSENBAUM, SHLITA

ASLAN SAFDEYE

MR. AND MRS. JACK SAFDEYE

DR. JOSEPH SAFDIEH

ARI SIPPER

MOSHE VINBERG

MATT WEINER

MICHAEL ZAMIR

MORRIS ZEITOUNE